WHAT WOULD ANIMALS SAY IF WE ASKED THE RIGHT QUESTIONS?

Cary Wolfe, Series Editor

(continued on page 251)

What Would Animals Say If We Asked the Right Questions?

VINCIANE DESPRET

translated by BRETT BUCHANAN

Foreword by Bruno Latour

posthumanities 38

UNIVERSITY OF MINNESOTA PRESS
Minneapolis · London

Cet ouvrage a bénéficié du soutien des Programmes d'aide à la publication de l'Institut Français. This work received support from the Institut Français through its publication program.

Originally published in French as *Que diraient les animaux, si... on leur posait les bonnes questions?* by Vinciane Despret. Copyright Éditions La Découverte, 2012.

Portions of "B for Beasts" were published in Vinciane Despret, "Il faudrait revoir la copie: L'imitation chez l'animal," in *Mimesis,* ed. Thierry Lenain and Danielle Loriès, 243–61 (Brussels: La Lettre Volée, 2007).

Translation copyright 2016 by the Regents of the University of Minnesota.

Published by the University of Minnesota Press
111 Third Avenue South, Suite 290
Minneapolis, MN 55401-2520
http://www.upress.umn.edu

Printed in the United States of America on acid-free paper

The University of Minnesota is an equal-opportunity educator and employer.

21 20 19 18 17 16   10 9 8 7 6 5 4 3 2 1

Library of Congress Cataloging-in-Publication Data
Despret, Vinciane, author.
What would animals say if we asked the right questions? / Vinciane Despret ; translated by Brett Buchanan.
Minneapolis : University of Minnesota Press, [2016] | Series: Posthumanities ; 38 |
Includes bibliographical references and index.
Identifiers: LCCN 2015036886| ISBN 978-0-8166-9237-8 (hc) |
ISBN 978-0-8166-9239-2 (pb)
Subjects: LCSH: Animal behavior—Miscellanea.
Classification: LCC QL751 .D44613 2016 | DDC 591.5—dc23
LC record available at http://lccn.loc.gov/2015036886

CONTENTS

FOREWORD

The Scientific Fables of an Empirical La Fontaine

Bruno Latour

Be prepared to read stories of "The Pig Who Tried to Lie," "The Much Too Clever Magpie," "The Elephant and the Mirror," "The Parrot Who Refuses to Parrot," "The Cow Who Wishes to Meditate," and "The Goats Who Cannot Be Counted," of "The Tick Who Believes She Is Charles Sanders Peirce" or of "The Penguin Who Has Read Too Many Queer Stories," and don't skip the one on "The Lemur and Its Ethologist Tried for Infanticide" and many, many others. Be prepared to read a lot of science but also to learn about the many ways to do good, bad, or terrible science. You are about to enter a new genre, that of scientific fables, by which I don't mean science fiction or false stories about science but, on the contrary, true ways of understanding how difficult it is to figure out what animals are up to. This is one of the precious books that pertain to the new rising domain of scientific humanities, meaning that to understand what animals have to say, all the resources of science *and* of the humanities have to be put to work.

The problem with animals is that everyone has some experience with them and tons of ideas on how they resemble humans, or not. So if you begin to offer disciplined accounts about their mores, you immediately run up against a stream of "but my cat does this," "I have seen on YouTube a lion doing that," "scientists have shown that dolphins can do this," "on my grandfather's farm pigs used to be able to do that," and so on and so forth. The good side of this is that whenever you mention animals, everybody is interested in what you have to say; the bad side is that your account will be drowned in alternative versions derived from totally different preoccupations and experiences of dealing with animals.

Most scientists, when faced with such a din of alternative accounts, will try to distance themselves from all of them, to start from scratch, and to mimic, as exactly as possible, what their fellow scientists in neighboring fields have done with physical objects and chemical reactions. Whatever ordinary people, pet owners, stockbreeders, conservationists, and TV documentarists have said will be pushed aside as so many mere "anecdotes." And the same will be done with what scientists from earlier centuries, or one's colleagues with a different training, claim to have noticed in some unusual circumstances, for instance, in their many field observations. "Enough anecdotes; let's start with real data in a controlled setting, the laboratory, to study the behavior of animals in as objective, disinterested, and distant a light as possible."

If the amateurs should be kicked out, those scientists claim, it is because they tell stories from which you will never know, when hearing them, if they speak of their emotions, attitudes, and mores or those of the animals *themselves*. Only the strictly controlled conditions of the laboratory will protect knowledge production from the pitfall of "anthropomorphism." Such a reaction produces an interesting paradox: only by creating the highly *artificial* conditions of laboratory experimentation will you be able to detect what animals are really up to when freed from any *artificial* imposition of human values and beliefs onto them. From then on, only one set of disciplined accounts of what animals do in those settings will count as real science. All other accounts will be qualified as "stories," and the storytellers will be dismissed as mere amateurs.

For the last twenty years, Vinciane Despret, trained in experimental and clinical psychology as well as in philosophy, has never stopped inquiring into this strange paradox: why is it that scientific knowledge about animals should be created under such artificial conditions to *get rid* of all the equally artificial situations in which humans encounter animals? Is the fight against anthropomorphism so important that it should give way to what she calls a generalized "academocentrism"? By this she means that only a tiny register of attitudes are imposed not only on animals but also on those reading scientific accounts. Is it not a little bizarre that naturalistic descriptions are supposed to be obtained by artifices, whereas the naturally occurring situations are considered a source of artificial fictions? Because knowledge, after all, is always produced for artificial reasons and in artificial settings, why not use the thousands of instances in which

humans interact "naturally" with animals—including the daily practices of
handling laboratory animals and of imagining new experimental designs,
as well as the practices of trainers and breeders—to *accumulate* knowledge,
instead of *subtracting* it?

Vinciane Despret belongs to a special breed of empirical philosophers.
It is sometimes overlooked that empiricists come in two main varieties: the
subtractive empiricists and the *additive* empiricists. The first are interested
in grounding their claims, but only on the condition that a claim decreases
the number of alternatives and limits the number of voices claiming to
participate in the conversation. What they are after is to simplify and to ac-
celerate, sometimes even to eliminate accounts altogether and, if possible,
silence storytellers as well. The *additive* empiricists are just as interested in
objective facts and grounded claims, but they like to add, to complicate,
to specify, and, whenever possible, to slow down and, above all, hesitate
so as to multiply the voices that can be heard. They are empiricists, but
in the fashion of William James: if they want nothing but what comes
from experience, they certainly don't want *less* than experience. As Isabelle
Stengers, one of the most important sources of inspiration for Despret's
original method, likes to say, science debases itself when it argues from
its successes to eliminate other accounts. Rather than purveyors of the
"either–or," Stengers and Despret are great proponents of "and–and."

How to be a consistent *additive* empiricist? By first taking very seriously
and reading very carefully all the *subtractive* empiricists' accounts. Despret's
genius is to read the scientific literature not to review it—that is, to extract
the few solid facts and dismiss the rest as irrelevant—but to explore what
it reveals about the endless difficulties of creating meaningful settings to
replicate some of the conditions where humans and animals interact or,
more important, where animals interact with other animals. And then, in
a second move, she uses such difficulties to shed light on how the many
other types of knowledge-producers also deal with animals, but do so by
taking a totally different sort of care. Laboratory-based accounts have to
be added, of course, the discoveries being as marvelously revealing as they
are, but without being granted the power to eliminate alternative accounts.

Such a generous attitude toward the scientific literature generates an
extraordinary effect, what I like to call the "Despret effect," by which an
austere body of science about hundreds of often bizarre experimental

situations becomes fascinating to read. It is treated with humor, but without any irony and, what is strangest of all, without any of the critical tone so often used by animal lovers against scientific claims. When you are an *additive* empiricist, it is all forms of subtraction that have to be resisted: eliminativism of those who wish to kick the amateurs out, but also eliminativism of those who dream of bypassing science altogether—two forms of competing and complementary obscurantism.

Thanks to the Despret effect, each time you react with indignation against an alternative version of what an animal is supposed to do is a new occasion on which to hesitate about how you ascribe agency to humans as well as to animals. You move from the question of anthropomorphism to the much more interesting one of *metamorphosis,* by which I mean not only to police the boundary between what is human and what is animal (a limited question if ever there was one) but to explore the protean nature of what it means to be "animated." Scientists, breeders, animal lovers, pet owners, zookeepers, meat eaters—we are all constantly trying to avoid deanimating or overanimating those beings with whom we constantly change shapes (*shape-changing* is the English equivalent of *metamorphosis*).

After numerous long-term inquiries, Vinciane Despret, in *What Would Animals Say If We Asked the Right Questions?* (magnificently translated by Brett Buchanan), has decided to present a great deal of her past work in a series of short chapters that read much like La Fontaine's fables, except that her fables are not grounded in a millenary-old folklore; each is instead grounded in a specific body of scientific and ethnographic literature about one or several animal encounters.

What relates this book to fables is of course that animals speak, or, more exactly, "would speak," *if only* we could ask the "right questions." Whereas in the traditional genre of the fable, there is no apparent problem in making the animal say something funny, critical, astute, ironic, or silly, here every instance of expression is related to *how* the questions are asked. And the questions are often funny, critical, clever, ironic, or downright silly—sometimes criminal (see the fable that could have been called "The Sadistic Harlow and His Monkeys"). So each fable brings us closer to what could be called the collective speech impairments of those who could make others say something if only they themselves were not so hard of hearing.

In Despret's hands, the ability to make animals say something relevant has a way of being infectious: silly questions create silly animals read by people who become even sillier; clever questions reveal clever animals able, through the transcription of their feats, to render readers more intelligent about the world. When reading Despret, there is no question that the world gains in complexity and that the meaning of what it is to be "animated" is deeply metamorphosed.

But what makes this book pertain to a renewed genre of scientific fables is that each short chapter ends *with a moral*—not the somewhat tedious moral lessons that La Fontaine liked to add to his stories but, on the contrary, a series of very audacious philosophical ones. In a way, Despret's fabliau is nothing less than a book on scientific methods that could be read not only by young scientists starting in the field of ethology but also by all those who are never sure how they should welcome the news coming from science about "their" animals.

In a way, the book can be read as a series of moral tales not only on how to do science but also, on behalf of the general public, on how to experiment on ourselves about our own ethical reactions. This is especially true of the question of how farm animals are treated—a very tricky issue. How can the question of agency, even in such a delicate case, be maintained through an additive and not a subtractive form of empiricism? In a fable that could be called "The Herder and the Laboring Cow," Despret mentions the study of her friend Jocelyne Porcher, whose particular stand is, "Of always thinking about humans and animals, farmers and their beasts, together. To no longer consider animals as victims is to think of a relation as capable of being other than an exploitative one; at the same time, it is to think a relation in which animals, because they are not natural or cultural idiots, actively implicate themselves, give, exchange, receive, and because it is not exploitative, farmers give, receive, exchange, and grow along with their animals."[1]

Why is it so difficult to avoid denying agency when dealing with animals? Well, because of this strange idea of always deanimating entities for fear of overanimating them, that is, of giving them some sort of "soul." What makes Despret's attempt so exceptional is her use of the very literature that tries to deanimate animals for the express purpose of showing how "animated" it is. But "animated" is as distant from having a soul as it is from acting as a computer. This she succeeds in doing not only with examples from the behaviorist turn—what has now become a treasure trove of funny anecdotes—but also with cases from the "sociobiological" turn, where genes had been endowed with so much causal agency that there was nothing left for the animals "acted" by their selfish genes to do "by themselves." Reductionism is in many ways an unachievable ideal as soon as you begin to foreground the experimental setup through which "reduction" is achieved. Interesting problems keep proliferating at every turn. This inner contradiction is never more visible than in the case of Lorenz. In a fable that could have been called "The Peacock and the Scientists," Despret writes, "Ethologists who follow [Lorenz's] approach will have learned to look at animals as limited to 'reactions' rather than seeing them as 'feeling and thinking' and to exclude all possibility of taking into consideration individual and subjective experience. Animals will lose what constituted an essential condition of the relationship, the possibility of *surprising* the one who asks questions of them. Everything becomes predictable. *Causes* are substituted for *reasons* for action, whether they are reasonable or fanciful, and the term *initiative* disappears in favor of *reaction*."[2] Except that Lorenz is also remembered as having renovated many of the earlier attitudes of attention and respect toward the surprising behavior of animals. So in the end, is Lorenz an additive or subtractive empiricist? Ah, if only Tschock the jackdaw could tell his side of the story!

For me the main reason why the moralities drawn, fable after fable, are so important for the scientific humanities and, more generally, for philosophy is that what Despret shows for mainly twentieth-century animals in their relations to humans is what had occurred in earlier centuries with physical, chemical, and biochemical entities. The endless number of relations humans had with materialities had been channeled into a much narrower set of connections with what comes to be called "matter." Materiality and matter are just as distinct sets of phenomena as a monkey studied in the field by Shirley Strum (see the fable "The Baboon and the Berkeley Lady") is from one seated in the chair of a behaviorist's laboratory in the 1970s. Except in this case, the exclusion of other voices, attitudes, skills, and habits is so much taken for granted now that we don't hear, nor can we imagine, that a huge operation has been going on to discipline agencies and, here too, to deanimate materiality rather forcefully so as to obtain, in the end, something like "a material world." And it is then into this highly simplified "material world" that poor animals—humans included—have to be inserted and asked to scratch a living.

But once you have been infected by Despret's generous lesson, you cannot stop extending the lesson elsewhere, for instance, into physics and chemistry. After all, it is Alfred North Whitehead, another major influence on her method, who has claimed that in physics too we should learn to become, again, additive and not subtractive empiricists: "For natural philosophy everything perceived is in nature. *We may not pick and choose.* For us the red glow of the sunset should be as much part of nature as are the molecules and electric waves by which men of science would explain the phenomenon. It is *for natural philosophy* to analyze how these various elements of nature are connected."[3] The great beauty of Despret's work is that she is indeed a "natural philosopher" thoroughly renovating not only the range of issues usually dealt with by philosophy but also the range of potential agencies with which "nature" is endowed. And in addition, she is doing this by stylistic inventions—the scientific fables—that mimic exactly, in their rhythm, their humor, their depth of knowledge about so many experimental settings, what we need to regain a connection with intelligent animals made to say intelligent things through the clever devices of scientists rendered intelligent by them—"them" being, well, every one of those thus assembled. Do the mental experiment: compare what wolves, monkeys, ravens, cows, sheep, dolphins, and horses were supposed to be able to do thirty years ago with the capacities with which they are endowed today; what has been opened up is an entirely new world of capacities.

The problem, and what makes Despret's work even more interesting, is that such an expansion of animal capacities has no parallel in what "human" agents are supposed to be able to do. This is where her work becomes significant for political philosophy. This is what Donna Haraway—another crucial influence on Despret's attitude—has done by offering the mutual relations established with her dog Cayenne as an example of the sort of attentions we would need to become again political agents. Deprived of the attention given to them by other "companion species," humans have lost the ability to behave as *humans.* This is what renders the fight against anthropomorphism so ironic: today most humans are not treated by sociologists or economists as generously as wolves, ravens, parrots, and apes are treated by their scientists. In other words, a book called *What Would Humans Say If They Were Asked the Right Questions?* remains to be written. What is sure is that, as of now, and at least in Vinciane Despret's sure hands, animals seem to be able to tell quite a lot of moral tales that would bring immense benefits if humans were allowed by *their* scientists to hear them.

ACKNOWLEDGMENTS

Thank you to Éric Bataray, Éric Burnand, Annie Cornet, Nicole Delouvroy, Michèle Galant, Serge Gutwirth, Donna Haraway, Jean-Marie Lemaire, Jules-Vincent Lemaire, Ginette Marchant, Marcos Mattéos-Diaz, Philippe Pignarre, Jocelyne Porcher, Olivier Servais, Lucienne Strivay, and François Thoreau.

And a special thanks to Laurence Bouquiaux, Isabelle Stengers, and Evelyne Van Poppel.

HOW TO USE THIS BOOK

This book is not a dictionary. But it can be handled like an abecedary. One may follow the alphabetical structure, if one likes to do things in order. But one may also begin with a question that is interesting and that whets the appetite. I hope that one will be surprised *not* to find what one is looking for or what one expects. One may open the book in the middle and trust one's fingers, curiosity, chance, or some other order, or amuse oneself at the whim of the references (☞) that are sprinkled throughout the text. There is no prescribed meaning or key to the reading.

TRANSLATOR'S NOTE

Translating the alphabetical structure of this abecedary has led to the occasional challenge, but nothing that couldn't be solved with a little creative flexibility. With Vinciane's grace, and at her suggestion, we've rearranged the sequence of four of the chapters, effectively swapping M and P and T and W from the original French publication. As noted in "How to Use This Book," however, the book is designed to be read in whatever order the reader wishes, so these modifications should not affect the overall coherence of the book.

The work of translation is an endless task and one that continues well after the final page has been printed. I have been aided by the invaluable support and friendship of Vinciane Despret, first and foremost, and by Matt Chrulew, both of whom read the manuscript from A to Z. Further assistance has been provided by Jeff Bussolini, Kelly Darling, and many friends and colleagues along the way. However, responsibility for this translation remains mine alone. My thanks to Bruno Latour for writing the foreword and to all of the wonderful people at the University of Minnesota Press, especially Erin Warholm, Doug Armato, and Cary Wolfe. This translation is dedicated to Fiona and Owen.

A
FOR ARTISTS
Stupid like a painter?

"Stupid like a painter." This French saying goes back at least to the time of Murger's *The Bohemians of the Latin Quarter,* around 1880, and was always used as a bit of a joke in conversations. But why must the artist be considered as less intelligent than your average joe?
—Marcel Duchamp, "Should the Artist Go to College?"

Can one paint with a brush attached to the end of one's tail? The famous painting *Coucher de Soleil sur l'Adriatique,* which was presented at the 1910 Independent Salon, offers an answer to this question. It was the work of Joachim-Raphaël Boronali, and it was his only painting. Boronali's real name was actually Lolo. He was a donkey.

These last few years, and owing to the influence of the circulation of their works on the Internet (☞ **YouTube**), many animals have revived an old debate: can they be granted the status of artist? The idea that animals can create or participate in works of art is not new—leaving aside Boronali, a rather mischievous experiment that didn't really strive to pose this question. It's nevertheless the case that animals, for some time now, have collaborated (for better, but often for worse) in the most diverse of spectacles, which have led some handlers to recognize them as artists in their own right (☞ **Exhibitionists**). If we stick to pictorial works of art alone, the candidates today are quite numerous, albeit hotly debated.

In the 1960s, Congo, a chimpanzee studied by the famous zoologist Desmond Morris, sparked a controversy with his abstract impressionist paintings. Congo, who died in 1964, has a follower in Jimmy, a chimpanzee who was so bored that his trainer had the idea of bringing him some paint and who today gives daily demonstrations at the Niteroi Zoological Garden (just across the bay from Rio de Janeiro). More famous than Jimmy, and certainly more active in the art market, is Cholla, the horse who paints abstract works of art with his mouth. And as for Tillamook Cheddar, we have a Jack Russell terrier who executes her performances in public thanks to a device that takes advantage of her breed as a rat catcher (and above all her nervous disposition): her owner covers a blank canvas with a layer of smooth carbon impregnated with colors, which the dog proceeds to attack with strokes of her nails and teeth. While the dog executes her work of art, a jazz orchestra accompanies the performance. At the end of a dozen relentless minutes—on the part of the dog—the owner recovers the canvas and unveils it. What appears is a figure that has been made from the nervous and concentrated scratches on one or two parts of the painting. Videos of these performances circulate around the Internet. It must be recognized, and without judging the outcome, that the question can be asked as to whether there is true intention in the production of the work of art.

But is this the right question?

More persuasive in this respect, at least at first glance, is the experiment conducted with some elephants in the north of Thailand. When Thai laws banned the transportation of lumber by elephant, elephants found themselves out of work. Incapable of returning to nature, they were welcomed in sanctuaries. Among the videos circulating on the Web, the most popular ones were filmed at Maetang Elephant Park, approximately fifty kilometers from the city of Chiang Mai. These videos show an elephant creating what the film directors call a self-portrait, and as it happens, it's a picture of a stylized elephant holding a flower in its trunk. What allows commentators to call this painting a "self-portrait" remains to be explained. Would an extraterrestrial assisting a man who is drawing from memory the portrait of a man also be inclined, in such a case, to speak of a self-portrait? In the case of our commentators, do their claims consist of a difficulty in recognizing individualities, or is this just an old reflex? I'd lean toward the reflex hypothesis. The fact that once an elephant paints an elephant it becomes automatically perceived as a self-portrait is probably tied to the strange conviction that all elephants are substitutable one for the other. The identity of animals is often reduced to their species membership.

On viewing the images of this elephant at work, one cannot help but be a bit troubled: the precision, the exactitude, and the sustained attention

of the animal to what he is doing all seem to combine the same conditions of what would be a form of artistic intentionality. But if one looks
a little further and interests oneself in how the device is assembled, one
can see that this work is the result of years of apprenticeship in which the
elephants first learned to draw via sketches made by humans, and that it is
these same learned sketches that the elephants tirelessly reproduce. When
one thinks about it, the opposite would have been surprising.

Desmond Morris was also interested in the case of the elephant painters.
Taking advantage of a trip in the south of Thailand, he decided to have a
closer look at them. The duration of his stay did not allow him to travel
north to the Chiang Mai sanctuary where the elephants had become
famous artists, but a similar spectacle could be seen at the Nong Nooch
Tropical Garden. Here is what he wrote upon viewing the performance:
"To most of the members of the audience, what they have seen appears
to be almost miraculous. Elephants must surely be almost human in intelligence if they can paint pictures of flowers and trees in this way. What
the audience overlooks are the actions of the mahouts as their animals
are at work."[1] If one watches carefully, he continues, one can see that
with every stroke painted by the elephant, the mahout has tugged on the
elephant's ear, up or down for vertical lines and to the sides for horizontal
ones. So, Morris concludes, "very sadly, the design the elephant is making
is not hers but his. There is no elephantine invention, no creativity, just
slavish copying."

Now that's what one calls a killjoy. It always surprises me to see the
zeal with which some scientists rush to play this role, and how they take
on, with such admirable heroism, the sad duty of bearing bad news—
unless, of course, it's a matter of manly pride by those who are unwilling
to let themselves be taken where everyone will be fooled. It's the joy, at
any rate, that is cast aside in this story, as is the case when scientists devote
themselves to the cause of this kind of truth that needs to open our eyes:
the recognizable scent of "it's nothing but . . ." is a sign of this crusade of
disenchantment. But this disenchantment would not be possible were it not
for the price of a grievous (and likely not very honest) misunderstanding
as to what enchants and what brings joy, for the misunderstanding rests
on the belief that people naively believe in miracles. In other words, one

could only be so easily disenchanted if one were mistaken as to what the enchantment was.

There is indeed something enchanting in the performances given to the public. The enchantment, however, is not where Morris situates it. Rather, there is something more like a certain grace, a grace that can be seen in the videos and in an even more noticeable manner if one has the chance to see the elephants live, a chance that I had a little while after writing the first draft of these pages.

This enchantment emerges from the careful attention of the animal, from each line traced by her trunk—soberly, precisely, and decidedly, and yet also hanging, at certain moments, suspended, in a few seconds of hesitation—offering a subtle blend of affirmation and reservation. The animal, we would say, is entirely at one with her work. But above all, this enchantment arises by the grace of the attunement between living beings. It belongs to the achievement of people and animals working together and who seem happy, I'd even say proud, to do so, and it is this grace that the public recognizes and applauds as enchanting. The fact that there might be some "training tricks," like when an elephant is given a sign as to what to paint, is not what is important to those who attend the spectacle. What interests people is that what is unfolding before them remains deliberately indeterminate and that doubt can be maintained, whether it's required or freely permitted. No single response has the power to sanction the meaning of what is happening, and this very uncertainty, which is similar to that which we witness in a display of magic, is part of what makes us sensitive to its grace and enchantment.

I don't want to get carried away, then, in the debate by claiming that in the Maetang spectacle, unlike the one at Nong Nooch, the mahouts do not touch the ears of their elephants—I'd be unable to assert this anyhow, had I not looked again at the photos I had taken. This matters no more to me than any other killjoy that retorts that there must have been some other trick, different from one sanctuary to the next, and that I obviously failed to notice. Should we perhaps be satisfied to say that the southern elephants, as opposed to those in the north, need us to caress their ears to paint? Or that some elephants paint with their ears, just as we say that elephants of the south, of the north, and even those in Africa listen through the planting of their feet?[2]

The sadness that Morris evokes with his "very sadly, the design the elephant is making is not hers" is a sadness whose generous emancipatory offer I refuse. Of course the elephant's design is not her own. Who doubted this? Whether it is a trick or slavish apprenticeship by which the elephant copies what she has been taught, we always return to the same problem, that of "acting for oneself." I have learned to be distrustful of the manner in which this problem has been posed. Over the course of my research, I have noticed that animals are suspected, much more rapidly than are humans, of lacking autonomy. Manifestations of this suspicion proliferate especially when it concerns actions, such as cultural behaviors, that have for a long time been considered as proper to man. Consider, for instance, the recent observation of a striking display of mourning among a group of chimpanzees in a Cameroon sanctuary while confronting the death of a particularly loved peer. Because this behavior was sparked by the initiative of trainers to show the body of the deceased to her kin, criticism went into full swing: it was not real mourning; the chimpanzees should have manifested it spontaneously, somehow "all by themselves" (☞ **Versions**). As if our own grief in the face of death arose all by itself, as though becoming a painter or artist did not come from learning the gestures of those who preceded us, such as the continuation, over and over again, of themes that were created before us and that each artist hands down.

Of course the problem is much more complicated than this. But the manner by which it is posed in terms of an "either . . . or . . ." offers no chance to complicate it or make it interesting. Among the situations to be considered, the work of art does not appear to be the act of a single being, whether human ("it all comes down to human intentions," as some affirm) or animal (it is the animal who is the real author[3] of the work of art). What we are dealing with are complicated *agencements*:[4] it consists in each case of a composition that "makes" an intentional *agencement,* an *agencement* inscribed in heterogeneous ecological networks that combine, to go back to the case of the elephants, sanctuaries, trainers, amazed tourists who take photos and circulate them on the Web or buy their works of art to bring back home, nongovernmental organizations who sell these works to help care for the elephants, the elephants who found themselves unemployed following the law that prohibited them from transporting lumber . . .

I cannot therefore bring myself to answer the question as to whether animals are artists, be it in a sense that is close to or far from our own (☞ **Oeuvres**). Instead, I would choose to speak of achievements. I would opt then for terms that have been proposed or imposed on my writing within these pages: beasts and humans accomplish a work together. And they do so with the grace and joy of the work to be done. If I let myself be called by these terms, it's because I have the feeling that they are able to make us sensitive to this grace and to each event they accomplish. Isn't this what matters in the end? To welcome new ways of speaking, describing, and narrating that allow us to respond, in a sensitive way, to these events?

B
FOR BEASTS
Do apes really ape?

For a long time, it has been difficult for animals not to be stupid [*bêtes*], or even very stupid. Of course, there have always been generous thinkers, amateur enthusiasts, and those who are stigmatized as unrepentant anthropomorphists. The literature today, in our time of rehabilitation, is pulling them out of their relative obscurity in the same way as it prepared the case of all of those who made the animal into a soulless machine. And this is a good thing. But even if it is helpful to strip down these dominant machine-like discourses that have rendered beasts stupid [*rendre bêtes les bêtes*], it would be instructive to interest ourselves in the little machinations and less explicit forms of denigration that present themselves under the often noble motives of skepticism, obeying the rules of scientific rigor, parsimony, objectivity, and so on. For instance, the well-known rule of Morgan's canon states that when an explanation that draws on lower psychological competencies is in competition with an explanation that privileges higher or more complex psychological competencies, the more simple explanation ought to prevail. This is but one way, among other much more subtle ways, of talking nonsense [*bêtifier*] and whose detection at times demands laborious attention, even an uncompromising suspicion that borders on paranoia.

The scientific controversies about whether competencies should be recognized in animals are the best places for beginning such detection. Those that deal with imitation in animals are exemplary in this regard. It is all the more interesting because it will eventually end, after a long history and reasonably turbulent controversy, with the rather bizarre question, *do apes ape?*[1]

History shows us that what is at stake in these conflicts about the attribution of sophisticated competencies to animals can often be read, if one may forgive this barbarism, in terms of "proprietary rights of properties"

7

[*droits de propriété de propriétés*]: that which belongs to us—our "ontological attributes," like laughter, self-consciousness, knowing that we are mortal, the prohibition of incest—must remain our own. But how do we get from this to the confiscation from animals of what has been attributed to them?! One might expect scientists to be particularly touchy on certain questions of rival competencies, especially when philosophers have already been the object of this accusation—it is said that they become completely irrational as soon as the question concerns whether animals have access to language. Could imitation be for the scientists what language is for the philosophers?

Another hypothesis, and one that is more empirically supported, could take into consideration the unfortunate predilection scientists have for what are called "deprivation experiments." With deprivation experiments, the question of "how do animals do this or that?" becomes instead "what must be removed such that they no longer do this or that?" This is what Konrad Lorenz called the model of breaking down.[2] What happens when we deprive a rat or ape of his eyes, his ears, of this or that part of his brain, of all social contact (☞ **Separations**)? Is he still able to run a labyrinth, or control himself, or have relations with others? The serious fondness for this kind of methodology likely contaminates, much more broadly, the habits of certain researchers and now takes the shape of a strange ontological amputation: that apes can no longer ape.

The story, however, did not exactly begin like this. The question of imitation entered the natural sciences when George Romanes, a student of Darwin's, returned to one of his mentor's observations. Darwin had noticed that bees that gathered pollen on a daily basis from the flowers of dwarf beans by feeding from the open corolla of the flower modified their behavior when bumblebees joined in with them. The bumblebees used an entirely different technique whereby they pierced little holes in the calyx of the flower to suck out the nectar. The very next day, the bees were feeding the same way. Though Darwin cites this example only in passing as evidence of shared capacities among humans and animals, Romanes raises another theoretical significance: imitation allows us to understand how, when an environment changes, one instinct can give way to another, which then spreads. This is a lovely theoretical turn inasmuch as the imitation turns out to be what provokes the break or variation; it makes an "other" with the "same." Up to that point, this story did not consider rival possibilities. But this bifurcation did not catch on, because Romanes will add a commentary. It is, he wrote, easier to imitate than to invent. Furthermore, even if he conceded that imitation is evidence of intelligence, it is nevertheless an intelligence of the second order. Of course, he added, this faculty depends on observational learning, and thus a more evolved animal will be more capable of imitating. But Romanes's concession will be tempered by another argument: as intelligence gradually develops with children, the faculty of imitation diminishes, to the extent that we can consider it inversely proportional "to originality or higher powers of the mind. Therefore," he continues, "among idiots below a certain grade (though of course not too low), it is usually very strong and retains its supremacy through life, while even among idiots of a higher grade, of the 'feeble-minded,' a tendency to undue imitation is a very constant peculiarity. The same thing is conspicuously observable in the case of many savages."[3] As one can see, the faculty of imitation, itself hierarchical, participates in the organization of a hierarchy of beings that goes far beyond the problem of animality.

The double form of hierarchy that Romanes proposed—the hierarchy of modes of learning and that of intelligent behaviors—continued on after him, but it was far too simple, especially to address this difficulty: how can one put on the same footing the "sheepish" behavior of sheep (who are faithful imitators with or without their Panurge) and the behavior of parrots (who were once thought to be brainless) and apes who ape?[4] One is thus distinguishing between an instinctive imitation and a reflexive imitation, between mimesis and intelligent imitation, and, to distinguish between birds and others, between vocal imitations and visual imitations. All naturalists are in agreement on the fact that vocal imitations require a level of intelligence that is far less advanced than visual imitations. The anthropocentric aspect of this hierarchy, which has been established by beings for whom vision is a privileged sense, remains an open question.

What is also being differentiated, in a parallel way, are the intentional and active educational processes of responding to a plan and the imitation at work in passive and involuntary learning. This distinction deserves to be questioned precisely because it is familiar and obvious to us. Imitation is not only the methodology of the poor but is inscribed within the major categories of Western thought, categories that themselves hierarchize the regimes of activity and passivity. These categories, we know, do not sum up the distribution of regimes of experience or behavior but rather hierarchize beings who will be preferentially attributed with these behaviors. The distinction initiated by Romanes, between real intelligence (as evidenced in intentional learning) and a poor intelligence, attains its decisive form in the valorization of *insight* in the research on chimpanzees by Wolfgang Köhler.[5] *Insight*, which can be translated as "comprehension" or "discernment," is the capacity that allows an animal to suddenly discover the solution to a problem without having to pass through a series of trials and errors, as is the case in learning by conditioning. More precisely, *insight* was not coined to differentiate it from imitation but rather constituted a weapon for a bastion of resistance against the impoverishment imposed by behaviorist theories, which saw the animal as no more than an automaton for whom understanding is limited to simple associations. These associations were meant to exhaust all explanations with respect to learning. Behaviorists, it's worth noting, could hardly be bothered with imitation, and for good reason: their dispositives were conceived to study, with little exception, the animal acting alone. Imitation would remain confined to the margins of animal psychology and ethology.

When imitation does interest researchers, it is defined as the expedient of the poor and as something that allows animals to simulate cognitive capacities that they in fact do not possess. It's a cheap "trick," for lack of anything better, a fake, an easy way out that gives the appearance of real competency. Imitation is the antithesis of creativity (one can see how it is the opposite of *insight*), even if it can appear to some as a shortcut to excellence and thus constitute evidence of a certain form of intelligence.

In the 1980s a radical change occurred. Under the combined influence of child developmental psychology and fieldwork, imitation becomes not only an interesting subject again but changes status. It is now seen as a cognitive capacity that not only requires complex intellectual capacities but, even more, is indicative of highly elaborate cognitive competencies.[6] On one hand, imitation requires that the imitator understand the other's behavior as a directed behavior comprising desires and beliefs. On the other hand, this exercise leads to even more noble faculties: first, the possibility of understanding that the other's intentions lead to the development of self-consciousness, and second, the mode of transmission that enables imitation would be a vehicle for cultural transmission.

In short, once self-consciousness and culture were implicated, the stakes became more serious. Imitation would from now on open the doors to the cognitive paradise of mentalists—those who are capable of thinking that what others have in mind is different from what is in their own mind and of making plausible hypotheses to this effect (☞ **Pretenders**)—and to the social pantheon of cultural beings.

What followed was thus entirely predictable. The promotion of imitation to the status of sophisticated intellectual competency was accompanied by an incredible number of proofs that animals, in fact, did not imitate or were incapable of learning by imitation. It is now that we return to our question, which is the title of a celebrated article: *Do apes ape?*[7] The controversy ignited. Two camps formed on either side of a line that is easily mapped: fieldwork researchers increased their observations testifying to imitation, while experimental psychologists struck them down with the support of experiments.

Proponents of the theory of imitation summon the observations of gorillas that pluck stems from trees in a very sophisticated manner.[8] Their technique is transmitted by imitation, and we can see resemblances taking shape among peers who eat together. Orangutans can also be called in for help. At rehabilitation centers where researchers observe their progressive return to nature, orangutans can be seen wiping up and washing after themselves, brushing their hair and teeth, attempting to light a fire, siphoning a jerrican of gas, even writing, albeit in an illegible way—incidentally, the orangutans seem to be oddly lacking in enthusiasm with respect to the idea of returning to nature. "These are only anecdotes," the experimentalists calmly respond. Or better still, each of these examples can receive a different interpretation if one follows Morgan's canon.

The famous titmice that, to the great displeasure of milkmen, opened bottles of milk left on the front steps of English houses during the 1950s, and in doing so disseminated their techniques through a mode that showed the resilience of imitation, were summoned to the laboratory. The fact that these very titmice were able to modify their strategies to match the milkmen who were themselves adopting different systems of closing their bottles, and that the titmice circulated each new technique step by step, was not going to deter the experimenters. It was thus up to the titmice to prove their true talent for imitation. Within an experimental group control, however, the titmice were easily unmasked: the titmice that were confronted with a bottle that had been previously opened without their assistance fared just as well as those that received an example of how to open it from a fellow titmouse. It was therefore not imitation but *emulation*. Experimenters have also summoned the apes, and the verdict is similarly irrevocable: it is not true imitation that we see but simple mechanical associations that resemble copycat behavior. In fact, it consists of a pseudo-imitation. And there you have it: apes *imitate imitation*. But obviously they do so without deceiving the researchers who are always on the lookout for counterfeits. Only humans truly imitate.

The experiments will continue to multiply in the laboratory to test this hypothesis, which is but a translation of a more general thesis: that of *the* difference between humans and animals. Humans are thus summoned, and for good measure, they'll stick with children, for it is they who currently hold the responsibility of being compared with chimpanzees. By the end of the experiments, apes were losing right across the board. The psychologist Michael Tomasello asked chimpanzees to observe a model retrieving food using a T-shaped rake. The chimpanzees went on to accomplish the task—but did so using a different technique. The verdict? Chimpanzees don't imitate because they cannot interpret the original behavior as a goal-oriented behavior. They do not understand the other as an intentional agent similar to themselves as intentional agents.

When confronted by an experiment with artificial fruit (a locked box in which fruit is found for nonhuman primates and candy for the small humans), children demonstrate touching loyalty to all of the experimenter's gestures, even going so far as to repeat the gestures several times. The chimpanzees also open the box without any problem, but without using the technique of the model or the important details of the operation. It is not imitation, but as was the case with the titmice, just *emulation*.

What can we say about this experiment, except what we already knew? That human children are more attentive to the expectations of human adults than chimpanzees . . .

Things become a bit more complicated, however, once Alexandra Horowitz has decided to revisit certain aspects of the problem. She will compare adult subjects with children—and the adult subjects are actually psychology students. The box is identical to the one used for the children, except that the adult subjects have a chocolate bar instead of candy. The experiment was a disaster. The students were just as bad as the chimpanzees inasmuch as they used their own techniques to open the box and showed no regard for what they had been shown, with some even closing the box afterward, which their model had not done. Horowitz laconically infers that the adults behaved more like chimpanzees than they did like children. Therefore, she concludes, if Tomasello were right, one would have to infer that adults do not have access to the intentions of others.[9]

Returning to what was being asked of the chimpanzees, it's interesting to understand how these dispositives "make stupid" [*rendent bête*]. One must pay attention to the blind spots that remain in these kinds of experiments. What this dispositive demonstrates is no more than the relative failure of these chimpanzees to conform to our manners or, better, to the cognitive habits of scientists. The scientists have not wanted to engage in the difficult work of following these living beings in their relations to the world and to others but have imposed on the chimpanzees their own without for a moment questioning how the chimpanzees interpret the situations that have been put to them (☞ *Umwelt*). It's actually quite astonishing to think that it is these same researchers who most vigorously denounce the anthropomorphism of their opponents when it comes to attributing to animals competencies that are similar to our own. And yet one can hardly imagine a more anthropomorphic dispositive than those that they have applied to apes!

These experiments, in short, cannot pretend to compare what they compare because they are not measuring the same things. In pretending to put imitative capacities to the test, the researchers were in fact attempting to produce docility. How else should we speak of this attempt to have them imitate our manner of imitation? And when they refused, it was noted as a failure on the part of the apes. The fact that children exaggerated their imitation should have been a clue: children understood the importance, for the researcher, of the fidelity of their actions. In this respect, the apes had a much less complacent and above all more pragmatic attitude; they were not pursuing the same goals.

Or might it be that the apes never imagined that the human suppliers of candy expected of them something as stupid as imitating, one gesture after another, without interruption? No doubt this is what animals ultimately lack: imagination.

C

FOR CORPOREAL

Is it all right to urinate in front of animals?

Nobody knows what the body can do, wrote Spinoza, the philosopher. I don't know if Spinoza would approve of the following elaboration, but it seems to me that a lovely experimental version for exploring this enigma can be found in the practices of some ethologists: "We did not know what our bodies were capable of; we have learned with our animals." Many female primatologists have remarked, for instance, that their fieldwork could affect the biological rhythm of their menstrual flow in very perceptible ways. To cite just one example, Janice Carter recounts that her menstrual cycle was completely thrown off while living with female chimpanzees that she was rehabilitating in the wild. Owing to the shock of her new living conditions, she experienced amenorrhea for six months, and when her cycle returned, it had an unexpected rhythm: during the years of fieldwork that followed, her cycle attuned itself to those of the female chimpanzees and became a thirty-five-day cycle.

References to the bodies of ethologists are nevertheless not very numerous; when they do appear, they are for the most part only briefly mentioned and usually in the form of a practical problem to be solved. And yet one finds in some of them, either explicitly or implicitly, a story in which their bodies will be actively mobilized in a particular form, namely, that of a mediating device [*un dispositif de médiation*].

One of the more explicit examples of this can be found in Donna Haraway's analysis of the fieldwork of the baboon primatologist Barbara Smuts. When Smuts first began her fieldwork at Tanzania's Gombe Stream National Park, she wanted to act as she had been instructed: so as to habituate the animals to one's presence, one has to learn to approach them gradually. To not unduly influence them, one must act as if one is invisible, as if one is not even there (☞ **Reaction**). As Haraway explains,

Smuts's actions consisted of being "like a rock, to be unavailable, so that eventually the baboons would go on about their business in nature as if data-collecting humankind were not present."[1] Good researchers were those who, by learning to be invisible, could observe the natural scene from close up, "as if through a peephole."[2] Practicing habituation by becoming invisible, however, is an extremely slow and arduous process and one that all primatologists agree is often doomed to fail. And if it is doomed to fail, it is so for one simple reason: because it is based on the idea that baboons will be indifferent to indifference. What Smuts could not ignore over the course of her efforts was that the baboons often watched her and that the more she ignored their gaze, the less they seemed satisfied. The only creature who believed in the so-called scientific neutrality of being invisible was Smuts herself, for ignoring the social cues of the baboons was anything but neutral. The baboons must have detected someone who was outside of every category—someone who gave the appearance of not being there—and had to wonder whether this being could be educated, or not, according to the criteria of being a polite guest among the baboons. In fact, it all came from the conception of animals guiding the research: the researcher is the one who poses the questions, and they are often a far cry from imagining that the animals themselves may be posing just as many questions of their own, and maybe even the same questions as the researcher! People can ask whether baboons are or are not social subjects without ever thinking that the baboons must also be asking the exact same question with respect to these strange creatures with such bizarre behavior. If the baboons ask themselves "are humans social beings?" the answer would obviously be no. And basing their actions on this answer, for example, the baboons avoid their observers or do not act normally, or even act quite strangely, because they are so thrown off by the situation. The way that Smuts resolved this problem is much easier to say than to do: she adopted a behavioral style similar to that of the baboons, adopted the same body language as them and, in short, learned what was and was not appropriate to do with the baboons. "I," Smuts writes, "in the process of gaining their trust, changed almost everything about me, including the way I walked and sat, the way I held my body, and the way I used my eyes and voice. I was learning a whole new way of being in the world—the way of the baboon."[3] She also borrowed from the baboons their way of addressing one another. As a result of this, she writes, when the baboons began shooting her evil looks that forced her to distance herself, this paradoxically constituted enormous progress: she was no longer treated as an object to be avoided but as a trusted subject with whom they could communicate, who would distance herself when signaled to do so, and with whom things could be clearly established.

Haraway connects this story with a more recent article in which Smuts evokes the rituals she and her dog, Bahati, create and assemble [*agencent*]; they produce, according to her, an embodied communication. It is an exemplary choreography, Haraway comments, of a relation of "respect," in the etymological sense of the term, namely, of "looking back," of learning to respond and to be respondent, to be responsible.[4]

But one could just as easily read this as that which draws an outline, both very empirical and speculative, of what the sociologist Gabriel Tarde calls an interphysiology, that is, a science of the *agencement* of bodies.[5] From this perspective, the body renews the Spinozist proposition: it becomes the site of what can affect and be affected, a site of transformations. Above all, and to underline what Smuts puts into play, it is the possibility of becoming not exactly the other through metamorphosis but *with the other*, not in the sense of feeling what the other is thinking or of feeling for the other like a burdensome empathizer but rather of receiving and creating the possibility to inscribe oneself in a relation of exchange and proximity that has nothing to do with identification. There is, in fact, a kind of "acting as if" that leads to a transformation of self, a deliberate artifact that cannot and does not want to pretend toward authenticity or to some kind of romantic fusion that is often evoked in human–animal relations.

We are, moreover, quite far removed from this romantic version of a peaceful encounter when Smuts insists on the fact that progress was clearly visible to her when the baboons began to make her realize that conflict was possible when they shot her evil looks. The possibility of conflict and of its negotiation is the very condition of the relation.

Still within the domain of baboon primatology, one can find in the writings of Shirley Strum a different variation of the use of the body. She recounts in her book *Almost Human* that one of the problems she encountered in the early days of her research was knowing what she could and could not do with her body in the presence of baboons.[6] This problem presented itself, for example, when Strum had to respond to an urgent need to urinate. The thought of leaving her spot to go hide behind her truck, which was parked far away, presented a real dilemma, for it is almost always certain (and I have heard a number of researchers express the same sentiments) that at the very moment when you absent yourself, it is then that something interesting and very rare happens. In the end, Strum decided, and not without fear, no longer to go behind her truck. She undressed herself with a good deal of precaution, keeping an eye out all around her. The baboons, she writes, were flabbergasted by the noise, for in fact they had never seen her eat, drink, or sleep before. The baboons, of course, knew a great deal about humans, but they never closely approached, and they probably believed, she suggests, that humans did not have any physical needs. They thus discovered that they do, and drew certain conclusions, for the next time they had no reaction.

One can only speculate from what Strum describes. Her success certainly stems from the quantity of her field studies, her work, the quality of her observations, her imagination, her sense for interpretations, and her capacity to connect events that do not seem to be otherwise connected. Her success stems just as much from the tactfulness she has always shown in the creation of the encounter with her animals, and to which the question that she poses attests: is it all right to urinate in front of baboons? But I can't help but think that her achievement—this amazing relation that she was able to create with them—perhaps stems just as much from what they discovered that very day: that she had, like them, a body. When we read what Shirley Strum and Bruno Latour have written on baboon societies and the complexity of their relations, this discovery could hardly have been insignificant for them.[7] Because baboons do not live in a material society, and because nothing is stable in their social relations, and because every little disruption of a relation affects all the others in unforeseeable ways, each baboon must constantly undertake the continuous work of negotiation and renegotiation to create and restore the web of alliances. The social task is a creative task, one that consists of the daily construction of a fragile social order and of continuously reinventing and restoring it. To do so, baboons have only their bodies at their disposal. What might appear to be anecdotal might, for the baboons, have constituted an event: this strange being of another species has, in a certain respect, a body similar to their own.

Does this interpretation hold up? Has Strum been "socialized" in Smuts's sense of the word, that is, become a social being in the eyes of the baboons by allowing them to see a body that is, in a certain respect, similar to their own? One can only speculate about this.

These two stories are reminiscent of another one recounted by the biologist Farley Mowat. This one, however, does not come from scientific literature, strictly speaking, and his writings have been quite controversial. Furthermore, it presents a series of serious reversals. On one hand, this story falls more within the category of a transgression of good manners than it does of a real desire to be an acceptable host. On the other hand, and regarding what Smuts recounts, it turns the question upside down: it is not the hosts who are required to be politely accountable as social beings but the observer.

Mowat's story begins at the end of the 1940s, when he was invited to lead an expedition to evaluate the effects of wolf predation on caribou populations.[8] The field would prove to be a harsh test, as Mowat spent long periods alone in his tent, in the middle of the territory of a pack of wolves that he was observing. Just as prescribed in the rules that Smuts evoked, Mowat also took great care in being as discreet as possible. As time went on, however, the biologist slowly but surely experienced the more and more difficult fact of being totally ignored by the wolves. He didn't exist. The wolves passed by his tent every day without showing the slightest bit of interest. Mowat thus started to consider a way that would oblige the wolves to recognize his existence. According to him, the method of the wolves imposed itself on him: he had to claim a right of ownership. And this is just what he did one night, taking advantage of when they departed to hunt. It took him all night long, and several liters of tea, but by dawn every tree, shrub, and tuft of grass that had been previously marked by the wolves had now been marked by him. Mowat now awaited the return of the pack, and not without concern. As usual, the wolves passed by his tent as if it did not exist, until one of them stopped in a state of total surprise. After a few minutes of hesitation, the wolf turned around, sat down, and fixed upon the observer with an uncanny intensity. Mowat, overwhelmed by anxiety, decided to turn his back on the wolf to signify that the wolf's insistence contravened the most elementary rules of manners. The wolf then started to systematically tour the field while leaving, with meticulous care, his own marks on top of those left by the human. From this moment on, Mowat writes, his enclave was ratified by the wolves, and each of them, wolf and human alike, regularly passed one behind the other to freshen up their marks, each on his side of the boundary.

Beyond these reversals, these stories fall within a very similar regime: one that characterizes situations in which beings learn either to ask that what matters to them be taken into account or to respond to such a demand. And that they learn to do so with another species. This is what gives such a remarkable and particular flavor to these scientific projects, wherein learning to know what is observed is subordinate to learning and, above all, to *recognizing one another.*

D

FOR DELINQUENTS

Can animals revolt?

On the beaches of St. Kitts in the Caribbean islands, humans and vervet monkeys share the sun, sand, . . . and rum cocktails. The term *share* no doubt translates more accurately the understanding that the monkeys seem to have of the situation than that of the humans, who, for their part, try as best they can to protect their drinks. And without much success; their rivals appear to be seriously motivated. In fact, this habit of vervets is well established. For more than three hundred years, these monkeys have been getting drunk, from the moment of their arrival on the island in the company of slaves who were themselves sent to work for the rum industry. The vervets developed a taste for rum by gleaning fermented sugarcane in the fields. Today, pilfering has replaced gleaning, and with these beings, humans are now facing an unprecedented expansion of what we have for a long time called *the* "social scourge."

But not all is lost, as these monkeys must have a lot to teach us about something, or solve one or another of our problems. The comments on the videos that document this story steer one way or the other.[1] A research program was launched by the Medical Council of Canada and the Behavioral Sciences Foundation of St. Kitts, where one thousand captive vervets were generously given various beverages. With statistics as evidence, the researchers concluded that the percentages of alcohol consumption by the monkeys align with those of humans. From one perspective, a good number of the monkeys seem to prefer juices and sodas and refuse the cocktails; of those that remain, on the other hand, 12 percent are moderate drinkers, whereas 5 percent drink themselves into total intoxication and are literally rolling under the tables. The females show less of a tendency to alcoholism, and when they do indulge in this miserable habit, they prefer the sweeter drinks. The behavior of the vervets while under the influence is distributed similarly to that of humans: some drinkers, in social

21

situations, are joyful and mischievous, others become morose, whereas others look to start fights. The moderate drinkers have habits that earned them, according to the researchers, the designation of "social drinkers": they prefer to drink between the hours of noon and 4:00 P.M. rather than in the morning. The drunkards, for their part, begin first thing in the morning and show a marked preference for alcohol that has been mixed with water as opposed to sweet drinks. Furthermore, if the researchers only provide access to alcohol during a reduced schedule of hours, they intoxicate themselves faster than you can say, right until they pass out. It is also observed that they monopolize the bottles and prevent the others from gaining access to them. All of this, they tell us, is a distribution of alcohol usage similar to our own. The researchers conclude that a genetic predisposition determines alcohol usage. And voilà, some good news. Finally we have an explanation that will free us of all of those details that uselessly complicate the situation, like cafés, weekends, the ends of the month and nightcaps, the fact of wanting to forget, of parties, loneliness, social misery, the last drink or two, the rum industry, the history of slavery, migrations and colonization, boredom from captivity, and many more still.

Returning to delinquency, examples of animals posing problems multiply almost everywhere. The heinous crimes can be amusing at times, at other times tragic. The baboons of Saudi Arabia have for a long time carved out a solid reputation as burglars who enter houses to raid the fridges. In terms of pickpockets, one can read in the July 4, 2011, issue of the *Guardian* that some crested black macaques in a national park in Indonesia stole the camera from photographer David Slater and only returned it to him after they had taken a good hundred pictures or so, mostly of themselves. As for racketeering and extortion, still in Indonesia, one learns that the macaques of Uluwatu Temple in Bali steal the cameras and bags from tourists and only return them in exchange for food. More generally, the thefts committed by animals in areas frequented by tourists have become innumerable, and they're accompanied, on some occasions, with aggression.

Much more dramatically, for a few years now, a rather sudden modification in the behavior of elephants has been observed. Some of them, for example, attacked a village in west Uganda and on several occasions prevented any passage by blocking off all roads. There has always been some conflict between humans and elephants, especially when space or food has been an object of competition, but this was not the case here; food was abundant, and there were not many elephants when these acts happened. In addition, similar cases have presented themselves elsewhere around Africa, and observers all mention that the elephants no longer behave like they did in the 1960s. Some scientists have evoked the emergence of a generation of "delinquent adolescents" owing to the effects of the deterioration of socialization processes that are normally accomplished within each herd; this deterioration is itself due to the last twenty years of intense poaching and even the elimination programs put in place by those responsible for wildlife management. These so-called removal programs have, in a number of herds, and according to a choice that remains questionable (as they surely all are), eliminated the eldest females, without realizing the catastrophic consequences this would have for the group. Other equally well-intentioned strategies to contend with local overpopulations consisted in displacing a few young elephants to reconstitute a herd elsewhere. They had similar results, for in such groups the matriarchs play an essential role. The matriarch is the memory of the community, the regulator of activities; she shares what she knows, but above all else, she is essential to the equilibrium of the group. When a herd encounters other elephants, the matriarch can know from their vocal signatures whether they are members of a larger or more distant clan, and she indicates how the encounter needs to be organized. Once her decision is taken and passed along to the members, the group calms down. Thus, of the herds that were reconstituted in a South African park at the turn of the 1970s, practically none of them survived. During autopsies, it was discovered that they had stomach ulcers and other lesions usually related to stress. In the absence of a matriarch, and left to themselves to provide a normal development and balance, the animals are unable to cope.

When the elephants began to attack humans without any apparent reason, these hypotheses were therefore considered: the elephants would have lost their points of reference and competencies that were previously provided during the long process of socialization among pachyderms. Some researchers further mentioned, in almost the same vein, that the elephants suffered from posttraumatic symptoms in a manner similar to humans.

This pathology rendered them incapable of handling their emotions, of coping with stress and controlling their violence. These hypotheses, as one can see, weave an increasingly tighter network of analogies with human behaviors.

In reading a recent book by Jason Hribal, yet another different version can be considered.[2] Hribal has been interested in what, in zoos and circuses, have for a long time been called "accidents" and that notably implicate elephants. These "accidents," in which animals attack, hurt, or kill human beings, in fact prove to be acts of revolt and, even more specifically, of *resistance* against the abuses they suffer. Hribal even takes it one step further: these acts in fact translate, behind their apparent brutality, a moral conscience on the part of the animals (☞ **Justice**).

Here again we see that the system of analogies feeds into the narratives. What was once the qualification of accidents is today the result of intentional acts, whose motives can be clarified and understood. Let us not forget what was concealed by the term *accident* in situations at circuses and zoos: in addition, of course, to the fact that this designation reassured the public about the exceptional character of the event, an accident defined all of the situations that do not call for true intentions. But one also called all those things claimed to be due to the instinct of the animal an "accident," which just as surely excluded the idea that the animal could have an intention or a motive (☞ **Fabricating Science**).

Hribal's proposal to translate "accidents" in terms of disapproval, indignation, revolt, or active resistance is nothing new. Though certainly more rare among scientists since the end of the nineteenth century, one still discovers this usage among "lay amateurs" such as trainers, breeders, caregivers, and zookeepers. This translation has nevertheless managed to impose itself on a recent situation that obviously left little doubt as to how it should be interpreted. It has been widely discussed ever since the start of 2009 when images were circulated on the Internet and some newspapers put it in the headlines. Santino, a chimpanzee at Furuvik Zoo north of Stockholm, took up the habit of bombarding visitors with stones as they passed nearby. Even more surprising, researchers who had become interested in this story noticed that Santino carefully planned his attacks. At the side of his enclosure, he gathered and hid stones close to the area where the tourists arrived, and he would do this in the morning before their arrival. In addition, he didn't do this on the days that the zoo was closed. When he ran out of material, he would manufacture his ammunition by working to loosen concrete rock available in his enclosure. According to the researchers, this was evidence of sophisticated cognitive capacities: the possibility of anticipation and, above all, future planning. There was little doubt that Santino put his skills to the service of expressing his disapproval.

The fact that chimpanzees use projectiles as weapons has already been observed in the encounters between groups. As further grist for the mill, they quite frequently collect their feces for their war plans—and though it's true that their feces are often the only weapon at this disposal within the confines of zoos, they do this just as much in the wild. It's thus that they sometimes welcome their strange conspecifics, that is, the unknown human. Quite a few researchers have learned this at their own expense.

Robert Musil once said of science that it transforms vices into virtues: that it seizes opportunities, employs cunning, considers the tiniest details for its own benefit, and cultivates the art of reversals and opportunistic new translations.[3] If there were a research program that merits this description, it would be the one led by William Hopkins and his colleagues. I'm not sure it is really necessary to add that it is proof of a remarkable devotion to the cause of knowledge; one can only appreciate its full scope by looking at the protocol and considering the duration of the experiment, which was close to twenty years.[4]

The question that guides Hopkins concerns "us"—one should note that it is rarely the opposite when examining chimpanzees in these experimental settings. The question participates in the large project to shed light on our origins as well as, a bit more modestly, the origin of some of the habits we have acquired over the course of evolution. In this case, the question is to retrace why right-handedness is privileged among most humans. A small detail. Except that many hypotheses have already been formulated on the basis of this one "detail." According to one of them—the one that interests Hopkins—the use of the right hand has developed with gestural communication. For instance, throwing toward a goal, and thus aiming at it, not only implicates the neural circuits responsible for intentional communicative behavior but also requires the ability to synchronize spatial and temporal information in precise ways. The gesture thus mobilizes the neural circuits that could prove to be essential to language acquisition. In other words, the practice of throwing could have constituted a determining factor in favor of the specialization of the left hemisphere with respect to communicative activities. So here then is the problem in which the chimpanzees are implicated: insofar as they are "just before us" on the evolutionary path, would they be right-handed? How can the chimpanzees be convinced to answer this question? One would have to work it out, simply on the basis of their tendency of wanting to throw. It did not escape the scientists that, from the first time that they encountered the chimpanzees, they indulged in the rather regrettable habit of throwing their feces.

It was regrettable, to be sure, until it became the royal road to knowledge. For almost twenty years now—twenty years!—researchers have devoted themselves with remarkable self-sacrifice to thrown excrement, by collecting as much information that they might one day shed light on one of the mysteries of humanization. Starting in 1993 with captive chimpanzees at the Yerkes National Primate Research Centre, the study enrolled, ten years later, chimpanzees from the University of Texas's M. D. Anderson Cancer Center; incidentally, when one knows what the chimpanzees had to submit to at this research center, one can imagine that the researchers' experimental proposal must have benefited from their approval.

Fifty-eight males and eighty-two females were observed throwing at least once, but only eighty-nine from among them were retained in the study, because the apes, for the strength of the results, had to have manifested this behavior at least six times. With a minimum score of six throws per chimpanzee over many years, one is obviously well beyond the setting of the first encounters, unless it is imagined that the researchers recruited an army of humans who accepted to play the role of the unfamiliar individual, which is not mentioned. Of course, the chimpanzees could use this method in other contexts, as in the case of disputes and when they wanted to draw the attention of another chimpanzee or an inattentive human. The scientists therefore had several strategies at their disposal. But another hypothesis can also be considered. The chimpanzees understood what the researchers expected from them and conducted themselves with grace, without being too rigid as to the rule of nonfamiliarity. Who knows from among all of the good reasons which ones may have aroused their motivation . . .

Two thousand four hundred and fifty-five throws were observed between 1993 and 2005—and these are only some of the ones for which the researchers were the target, because not included in these results are the attempts made by the less consistent apes, the occasional throwers.

It was probably all worth the effort, for the results are conclusive: chimpanzees are, in this case, by and large right-handed.

E

FOR EXHIBITIONISTS

Do animals see themselves as we see them?

In a magnificent article titled "The Case of the Disobedient Orangutans," Vicky Hearne—the philosopher, and dog and horse trainer—recounts how Bobby Berosini responded when she asked him the question of knowing what motivated his orangutans to work: "We are comedians. *We* are comedians. Do you understand me?"[1]

To begin with, there is the "we." It's true that the very form of Berosini's show could have encouraged its usage, because the mise-en-scène never ceases to blur the roles and identities. At the beginning of the show, Berosini recounts how people often ask him how he is able to get the orangutans to do the things they do. He explains that he answers this question by asserting, "You have to show them who is boss." He offers to give a demonstration. Berosini brings out an orangutan named Rusty and asks him to jump onto a stool. Rusty looks at him with signs of utter incomprehension. Berosini explains it to him with a number of gestures. Rusty takes on a look of greater and greater perplexity. Finally, Berosini decides to show him, jumps up onto the stool . . . and the orangutan invites the audience to applaud the human. The show thus proceeds through a series of reversals and inversions, especially when the orangutans persist in refusing cookies that are offered, and by trying to distribute them to the audience, after each of their achievements, which compels Berosini to eat them himself.

"We are comedians." There are multiple ways to construct a "we," and we never stop experiencing it—and failing at it—every day.[2] How to understand this "we" that seems authorized by, or is the outcome of, Berosini's achievements? One might first consider that the relation of domestication is a privileged condition for the acquisition of this shared competence. But though this hypothesis would be pertinent in other contexts, it does not really apply here. Domestication implies that humans and animals are mutually transformed over the lengthy process that produces human domesticators and domesticated animals. But the orangutans are not domesticated animals. The term *wild* does not seem any more appropriate. But they could, with their trainer Berosini, subscribe to the role of *companion species,* to invoke Donna Haraway's beautiful expression, and they could even give a new inflection to the etymology upon which she builds the meaning of their relation: they are not only species *cum-panis,* species that share their bread; they are also species that *earn* their bread together.[3] The "we" that unites them could thus be constituted in the fact of "doing things" together (☞ **Work**). This is probably the case.

But the situation of Berosini and his orangutans presents an additional dimension. The work that unites them is not any old kind of work. It is a work of spectacularization and exhibition. What Berosini dramatizes and creates in the show arises, therefore, owing to a particular figure of this possibility of saying "we," that which is produced by the particular experience of the exhibit: the possibility of exchanging perspectives.

We need to slow down here. To start with, what I attribute to be a characteristic specific to the exhibit *in general* might be no more than a consequence of the kind of scenario Berosini has chosen. The spectacle of disobedient orangutans radicalizes this experiment of exchanges of perspective because all of the protagonists are endlessly invited, over the course of the gags and role inversions, to adopt the position of the other—the apes assume the role of the trainer, the trainer finds himself in the position of the animals, and we no longer really know who controls whom. Everyone plays along in the game—which is fabulated, and explicitly fabulated—of experiencing the point of view of the other by literally putting himself, as the English say, *in the other's shoes.* But can it not be imagined that, with this scenario, Berosini only pushed to its limits one of the possibilities of the very experience of the exhibit, which would be the capacity to adopt the point of view of one or an other—the perspective of the one whom one pretends to be, of the one for whom one does so, and of the one who asks you to do it?

Then, much more problematically, it is evident that many animals—the majority even—who are brought out to be shown in zoos and circuses are living, every day, the tragic experience of the separation between "them" and "us" (☞ **Delinquents**; ☞ **Hierarchies**). It's because they are animals, and not humans, that they are thus exhibited, enclosed, given as fodder for our gaze and forced to execute a number of things that obviously have no interest at all for them and make them miserable. In these stories, there is neither a "we" nor (even less so) the possibility of exchanging perspectives—if we were actively capable of this, animals wouldn't be where they are. I agree, but I would not want us to forget those situations that are, from another angle, likely more exceptional in making these events possible: situations in which a "we" is created and at the heart of which perspectives are exchanged. They are recognizable in the reversal of the consequences that I have just mentioned in passing: wherein animals discover interests and are obviously "going about their business," which is another way of saying that they are happy in a way that can't be too far from what we call "being happy."

What allows a situation of exhibition to encourage perspectival exchanges and lead to this possibility of constructing a "we," though admittedly partial, local, and always provisional? While reading the accounts of breeders, trainers, and people who practice agility with their animals, it was clear to me that the spectacularization would induce, arouse, or appeal to a particular skill: *that of imagining being able to see with the eyes of an other.*[4] It's notable that this possibility covers a narrow definition of perspectivism that indicates our ways of considering the relations toward the world and others. Other traditions have created different features and, especially if we follow the work of the anthropologist Eduardo Viveiros de Castro among the Amerindians, a form according to which animals would perceive themselves as humans perceive themselves: the jaguar perceives itself like a human does, so that, for example, what we call the "blood" of its prey it sees as manioc beer, and what we consider as its fur it perceives as clothing.[5]

Considering animals as having perspectives in this restricted sense, however, opens an entirely different access to the old problem of what is called "mentalism." Mentalist animals are those who are able to attribute intentions to others (☞ **Beasts**; ☞ **Pretenders**). This competence, scientists agree, rests on another one: that of self-consciousness. Self-consciousness, still according to these scientists, can be credited on the basis of a test, that of recognizing oneself in a mirror (☞ **Magpies**). To summarize, therefore, the animals who can recognize themselves in a mirror can be recognized (this time by the scientists) as having consciousness of themselves. They can therefore participate in the next test that will demonstrate the mastering of a hierarchically superior competence: understanding that what others have in their heads is not the same thing as what they have in theirs. They can then, on this basis, surmise the intentions, beliefs, and desires of others.

Even if I can admire the ingenuity, patience, and talent of the researchers who set up these mirror devices, I have always remained a bit puzzled by the rather exclusive privilege of the chosen test. Admittedly, it's rather interesting to succeed in interesting animals in what interests us: ourselves. And yet, on one hand, isn't this "we" that I have just asserted posed a little too lightly? Do mirrors interest all of us? Or might it be a particular manner of defining a relation with oneself in a tradition preoccupied with introspection and self-knowledge and haunted by reflexivity? More generally, on the other hand, it is not only that the mirror arises from an essentially visual problem but that it assumes that to know oneself is to recognize oneself, therefore posing the problem in a solipsistic fashion. It is with one's self alone that being a self is negotiated, in a specular way. The fact still remains that this mirror test has been imposed with such self-evidence that it has become decisive in the matter. But those who have been "excluded" from this test, those for whom the mirror has no significance or interest, shouldn't they be reconsidered according to other modalities?

The question I am posing to the exhibit allows us to revisit this possibility, for when an exhibit can arouse, grant, lead, or bring into existence a particular form of perspectivism, it seems to me much better situated to define (and distribute in a less parsimonious manner) a certain dimension of self-consciousness, no longer as a cognitive process but as an interrelational process.

This competence is visible in the exact complement to the capacity of thinking of oneself that is like showing oneself to another, thus to see oneself as others see you. In other words, the act of hiding oneself is complementary to, and not contrary to (as one might otherwise believe), the act of exhibiting oneself. For both are surely the same competence, namely, that competence that one must speak of when an animal hides himself knowing full well that he is hiding himself: *he knows how to see himself as others see him,* and it is what allows him to imagine or predict the effectiveness of hiding himself. Hiding oneself while knowing that one is hiding indicates, in other words, the implementation of a process consisting of the possibility of adopting the perspective of the other: "From the spot where he is, he cannot see me." An animal that hides while knowing that he is hiding is therefore an animal endowed with the possibility of perspectivism; it is an animal that shows this fact in an even more sophisticated manner, because one is no longer in the disjunction of seeing–not-seeing but in a declension of possibilities of what is seen: it plays on effects (☞ **Oeuvres**).

Let's return to the exhibit as a situation in which perspectival competencies are carried out. On what basis does one recognize that an animal actively exhibits and implements this accomplishment? My answer may be a surprise: *inasmuch as the one who works with animals describes it as such.* This is what results, in particular, from a reading of Vicki Hearne's writings, wherein she speaks of the work of trainers; or in another way, from the survey of animal breeders that I conducted with Jocelyne Porcher.[6] Over the course of the latter, we actually noticed that the themes of the beauty contests in which the animals participated engaged, on the part of the trainers whom we questioned, in a descriptive system that was not only clearly perspectival but was so in a way that seemed to approach the sense given it by Viveiros de Castro.

Indeed, in these situations, the breeders see their animals as being capable of seeing themselves just as *we would see ourselves if we were in their position.* Some of them, like Acácio and António Moura, two Portuguese breeders, do not hesitate to assert that their cow, by the end of the contest, "concludes with the belief that she really is different and particular." A bit more sternly, Acácio added, "They will perhaps end up believing that they are beautiful, that they are divas." Or again, according to some Belgian and French breeders, "I had a bull that participated in some shows, and he knew that he had to be handsome because when you took a photo, he immediately raised his head a little. It was like he posed, you see, just like a star!" Both Bernard Stephany and Paul Marty further confirm this, that the animal knows and actively participates in its own staging: "This cow was a star and behaved as if she was a star, as if she was a human person who participated in a fashion show, and that made an impression on us. . . .

On the podium, the cow looked about, she was positioned in relation to the stands, she was like this, and over there were photographers as well. She looked at the photographers, and then very slowly, while the people applauded, she turned her head and looked at the people who were applauding. . . . Right there, it was as if she understood that she was supposed to do that. Besides, it was magnificent, because it was natural."

What these breeders related—and I also heard this from dog trainers—can be said in a few words: animals and people have succeeded in becoming attuned to what matters to the other, to act so that what matters to the other also matters to oneself.

I know these accounts will elicit a few giggles. But these snickers will only prolong the long history through which scientists have obstinately disqualified the knowledge of their rivals in matters of animal expertise, namely, those of amateurs, breeders, and trainers, and their anecdotes and hopeless anthropomorphisms (☞ **Fabricating Science**). Such snickering furthermore validates the awkwardness with which I myself posed the problem when I maintained that we recognize a situation of active and perspectival exhibition *inasmuch as the one who works with animals describes it as such.*

It's true that one rarely finds laboratory scientists who credit their animals with a will to actively show that, yes, they know very well how to do what has been asked of them and that they want to do it. It's rare for good reason, for if experimental psychologists considered this, they would be forced to concede that animals are not simply in the process of "reacting" or being conditioned but that they exhibit what they are capable of because we have asked it of them (☞ **Reaction**). In most laboratories, we show something à propos of animals; the animals show us nothing. This is why conditioning experiments, for example, pertain to the regime of demonstration but not to one of spectacle. This is also why there are no subjects with perspectives within this type of experiment.

This is precisely what Berosini mocks with his orangutans who redistribute the cookies. His parody of conditioning, which turns itself against him, reopens the question of reinforcement as a motive, for the reward of food within the conditioning device has the effect of definitively closing off the question "why do they do that?" The reward, in short, considerably precludes any possibility of perspective by eliminating the specter of complicated explanations, such as explanations that would force one to take into consideration reasons for which the animal might take an interest in what has been asked of him (☞ **Laboratory**). To put it another way, the reward of food is a motive that can pull the rug out from under perspective.

By maintaining that one can recognize an exhibitionist or perspectival situation because someone who works with animals describes it as such, I am not at all inviting us to think of all of this as but a matter of subjectivity or interpretation. For the act of description not only translates an engagement of the one who offers this description but engages and modifies those who allow themselves to be engaged by it and whom the description *attunes* in a novel register. In this sense, what I designate as "description" corresponds to an offer that had been welcomed and that can, from now on, qualify the achievement of this welcoming.[7]

Laboratories might perhaps acquire more interest if scientists considered them as places of exhibition. They would consequently renew a literal definition of the public dimension of scientific practices (this dimension is generally assured by the publication of articles) and would confer on them at the same time an aesthetic dimension. In place of routine and repetitive protocols, scientists could instead substitute inventive tests through which the animals could *show what they are capable of* when we take the trouble of giving them propositions that are likely to interest them. The researchers would explore new questions that would have no meaning other than to be welcomed by those to whom the propositions are made. Each experiment, then, would become a true performance and would require tact, imagination, consideration, and attention—all qualities of good trainers and perhaps artists as well (☞ **Oeuvre**).

By using the conditional, as I have just done, I may have given the impression that these laboratories are still to be invented. But they do exist, and a few of them can be found in this abecedary. Some of them even fit this description quite well, though I can't guarantee that their scientists would recognize themselves in that depiction. But recall that this is precisely the status that I gave to descriptions: propositions always at the risk of the welcome that they will receive.

F

FOR FABRICATING SCIENCE

Do animals have a sense of prestige?

The behavior of peacocks has not yet awoken much interest from scientists, who are preoccupied more by their tails than by their social manners or cognitive competencies. The peacock has no doubt had a role in this and imposed its own preoccupations on researchers. In addition to the problems of physics relative to how the capture of light produces such iridescent colors, the tail has sparked much debate: how did evolution permit such a cumbersome ornament that, all told, ought to have seriously handicapped its owner? This is what one calls a paradox of evolution. Darwin, who never doubted an aesthetic sense among animals, will say that the males who present the most beautiful finery would be privileged by the females and would thus transfer this characteristic to their descendants. More prosaically, researchers who came after him will reject the idea that these attributes, no matter how beautiful they are, could arouse some aesthetic emotion. To the degree that the tail must have some utility, however, they considered that the exuberance informs the females as to the vigor and good health of the tails' owners (☞ **Necessity**).

The Israeli ethologist Amotz Zahavi will take up this problem in a different way by shifting the focus a bit. We need to get past the idea, he says, that this so cumbersome tail is truly a handicap.[1] It must certainly present a burden that increases a peacock's visibility to its predators and must seriously compromise its ability to flee. But if a male who is endowed with an impressive tail, and thus clearly handicapped, has survived, then it is because he has had the means to do so. And if the females are sensible, they will thus have a strong interest in choosing an individual who is really handicapped as the father of their offspring—to the effect that, to solve a paradox, there's not another one like it. In other words, a handicap as remarkable as a vibrant tail is a form of reliable and unambiguous propaganda for its recipients.

But it hasn't escaped some observers that there happen to be times when a peacock is hardly selective in its choice of recipients. To this end, Darwin relates the following strange scene: that of a peacock striving to fan its tail in front of a pig. His commentary falls in keeping with his conviction that there is an aesthetic sense among animals: males adore showing off their beauty, as the bird clearly requires any old spectator, be it a peafowl, turkey, or pig.

This type of hypothesis will completely disappear from the scene of natural history in the years following Darwin. And when the same observation is rediscovered from the pen of Konrad Lorenz, the founder of ethology, a completely different interpretation is required. The peacock's display of his tail is defined as an innate pattern of actions associated with specific internal energies.[2] Stated more clearly, this behavior is innate and fits within a sequence of actions and reactions that succeed one another according to a programmed order. The animal, subject to specific internal energies, enters a stage of appetitive desires: instinctively he sets out in search of an object that, once found, will trigger an "innate releasing mechanism" of stereotypical behaviors. In the absence of an appropriate stimulus, this energy accumulates until it finally "erupts" (the peacock displays its tail) in vacuo—in vacuo here designating the pig.

The sociologist Eileen Crist invites us to pay attention to this model and, above all, to the contrast between the two interpretations.[3] On one hand, with Darwin, one has an animal who is entirely the author of his escapades and who has an impression of his own beauty, with motives and intentions to this effect—an animal that initiates things, indeed even strays a little bit, and that at any rate leaves us open to surprise. On the other hand, we find a biological machine at the whim of uncontrollable laws and whose motivations can be mapped like a quasi-autonomous plumbing system. The animal is "impelled" by forces that, admittedly internal, he has no control over. The difference between the two descriptions seems to be modeled after that which the Estonian naturalist Jakob von Uexküll (☞ **Umwelt**) identified between a sea urchin and a dog: when a sea urchin moves around, it is the legs that move it, whereas when a dog moves around, it is the dog that moves the legs.

The contrast between Darwin and Lorenz can be expanded, for it is not specific to these authors alone. One notices that the naturalists of the nineteenth century show with respect to animals a generosity in their attributions of subjectivity, which has subsequently been qualified as unbridled anthropomorphism. The majority of texts by naturalists of this era abound with stories that credit animals with feelings, intentions, wills, desires, and cognitive competences. In the twentieth century, these stories are found confined to the writings and accounts of nonscientists: "amateurs," naturalists, caregivers, trainers, breeders, hunters. Amongst scientists, the discourse will be marked primarily by the rejection of anecdotes and the exclusion of any form of anthropomorphism.

The contrast surrounding animals, as it appears between the practices of scientists and those of nonscientists, is thus relatively recent. It was constructed in two time periods and in two areas of research. The first is situated at the turn of the twentieth century as psychologists specializing in animals brought them into the laboratory and did their best to get rid of these nebulous explanations of will, mental or affective states, or even that animals might have a view on the situation and interpret it (☞ **Laboratory**).

The second period is constituted a little later, mainly with Konrad Lorenz. It's true that the image one retains of Lorenz is that of a scientist who adopts his animals, swims with his geese and ducks, and speaks with his jackdaws. This image is faithful to his practice but less so to his theoretical work. On the basis of Lorenz's theoretical propositions, ethology will engage in a resolutely scientific approach: ethologists who follow his approach will have learned to look at animals as limited to "reactions" rather than seeing them as "feeling and thinking" and to exclude all possibility of taking into consideration individual and subjective experience. Animals will lose what constituted an essential condition of the relationship, the possibility of *surprising* the one who asks questions of them. Everything becomes predictable.[4] Causes are substituted for reasons for action, whether they are reasonable or fanciful, and the term *initiative* disappears in favor of *reaction* (☞ **Reaction**).

How is it that Lorenz can be credited, and rightly so, with a practice that has its basis in—and resulted in—these wonderful stories of domestication and surprises but at the same time be at the origin of an ethology that is so arid and so mechanical?

Part of the answer can be found by revisiting the moment when ethology was constituted as an autonomous scientific discipline. Lorenz wanted to create an academic and scientific discipline in which only those who followed the curriculum could claim competency in ethology. However, other nonacademic people could also legitimately declare themselves competent in the area. These are the "amateurs," hunters, breeders, trainers, caregivers, and naturalists whose practices are close to ethology and who know animals very well but who don't have any real theory. To legitimize the area of knowledge that he is attempting to constitute, Lorenz will "scientize" the knowledge of animals. Ethology becomes a "biology" of behavior, hence the importance of instincts, invariant determinisms, and innate physiological mechanisms that are explicable in terms of causes. This differentiation proves to be all the more imperative because the proximity with its rival is strong and experienced as especially more dangerous insofar as a good part of its scientific knowledge is fed primarily by the knowledge of amateurs.[5] In short, it was a matter of *removing the animal from common knowledge.*

Lorenz's successors will faithfully follow the program thus instituted. The strategy of "doing science," as a procedure of placing at a distance those who might claim to know (and how to know), will gradually translate itself into a series of rules. Thus the rejection of anecdotes (that so remarkably punctuate the discourses of amateurs) and above all the manic suspicion with regard to anthropocentrism appear as the mark of a true science. Scientists who inherit this history now manifest an intense distrust of any attribution of motives to animals—and all the more so if the motives are complicated or, worse still, resemble those that a human might have in similar circumstances. In this context, instinct is the perfect cause: it escapes from all subjective explanations, and it is at once a biological cause and motive (a motive, moreover, that completely escapes the knowledge of the subject himself). One couldn't dream of a better object.

This means, then, that the accusation of anthropomorphism does not really apply, or not always, to the act of attributing human competences to the animal but instead incriminates the procedure through which this attribution is carried out. Before qualifying any cognitive procedure, the accusation of anthropocentrism, in other words, is a political accusation, a "politics of science" [*politique scientifique*] that aims above all to disqualify a mode of thinking or knowing from which the scientific practice has tried to free itself, namely, that of the amateur.

This hypothesis invites us to revisit the situations where accusations of anthropomorphism are found to pose different questions about them. Who claims to be protected with this accusation? The animal to whom we ascribe too much, or too little, and thus fail to recognize its ways of being (☞ *Umwelt*)? Or does it consist of defending some positions, some ways of doing science, some professional identities?

To support, and complicate, the possibility of this second hypothesis, I suggest we reconsider the example of Amotz Zahavi, the Israeli etholo-gist whom I mentioned earlier for his contribution to the mystery of the peacock's extravagant tail.[6] Zahavi does not in fact work with peacocks but with very specific birds, the Arabian babblers. He has been observing babblers for more than fifty years in the desert of the Negev, and it is with them that he came to elaborate the handicap principle from which, in ad-dition to peacocks, many animals that show extravagant and ostentatious behavior have since benefited, that has since been applied to, in addition to peacocks, many animals that show extravagant and ostentatious behavior. The handicap principle states that some animals assert their value (their superiority, says Zahavi) in competitive situations by exhibiting a costly behavior. Recall that being adorned with attributes that make you an easy target to predators is a costly behavior, a handicap; if you have survived, it is because you have had the means to do so.

Babblers are rather cryptic birds; their handicap does not have to do with their appearance but with their everyday activities. According to Zahavi, they never stop exhibiting costly acts that allow them to win a bit of prestige in the eyes of their companions. For prestige is important within the community of babblers. It allows a babbler to reach coveted positions within the hierarchy, which especially means, within groups where in principle only one couple reproduces, the possibility of estab-lishing one's candidacy as a breeder. Costly and prestigious acts can take many forms: babblers offer gifts in the form of food; they voluntarily offer themselves as sentinels; they feed, without any apparent benefit to themselves, the nest of the couple that reproduces; and they can show remarkable courage in taking risks in fights with other groups or when a predator threatens one of their own. Of course, birds that feed a nest that is not their own are not rare, especially in subtropical species; ethologists have copiously documented these situations. The fact of ganging up against an enemy is not exceptional either. The presents, on the other hand, are less frequent, at least outside of their coupling relations. But the babblers do not conduct themselves the way other birds do. On one hand, they do this with an explicitly exhibitionist willingness. They want to be noticed by others and signal each of their activities by a distinctively coded little whistle. On the other hand, they bitterly dispute the right to do so. If an individual of a less elevated rank attempts to offer a gift to another of a more superior rank, it will have to undergo an unpleasant fifteen minutes, a very unpleasant fifteen minutes. Numerous observations have thus led Zahavi to think that the babblers have invented an original answer to the problem of competition within groups for which cooperation is a vital necessity: they are in competition for the right to help and to give.

I have had the chance to accompany Zahavi for a while in the field, and I have learned, with him, to observe and to attempt to understand the behavior of these amazing birds. I was just as interested in the way he himself observed, in the way that he constructed his hypotheses, de-crypted the signs, and made sense of the acts. During the same period as Zahavi, another ethologist, Jonathan Wright, was carrying out his own research on babblers. Wright is an Oxford-educated zoologist, one who adheres to the theoretical postulates of sociobiology. From this perspec-tive, babblers do not assist one another for reasons of prestige, as Zahavi maintains, but because they are programmed by natural selection to act in the way that best guarantees the longevity of their genes. Based on the fact that babblers from the same group would be related, this theory claims that helping at the nest is one way to favor one's own genetic patrimony because there is a high probability that the nest is composed of brothers, sisters, nephews, or nieces, whose bodies would be the vehicles of a part of this same patrimony.

In terms of methods of field research, Zahavi and Wright are at op-posite ends of the spectrum. Zahavi was educated as a zoologist, but for a long time now his practice has been secondary to the project of the conservation of babblers, which is closer rather to the practices of naturalists. In observing him, I could not help but associate his methods with those of anthropologists. What defined a sequence of observations began with a sort of greeting ritual. Because the territories of each group of babblers are so large, one never knows where they will be found. It's therefore simpler to call them, and this is what Zahavi does: he whistles and waits. And the babblers arrive. Zahavi greets them with offerings of breadcrumbs. Then, from the point of view of his procedure for reading their behaviors, he builds his explanations (what are they doing, and why are they doing that?) based on reasoning by analogy: "And if I were the babbler, what would I do? What would make me act in such a way?" Wright clearly shows his disagreement with this sort of procedure. One cannot claim anything if there aren't any experiments, for this is the requirement of a truly objective science. One must show proof, and to prove, one must experiment. According to him, the Zahavi's interpretive method clearly belongs to an anthropomorphic and anecdotal practice— where it is understood that an anecdote is generally defined, in this area, as an *uncontrolled* observation; that is to say, it is not accompanied by the "right" interpretive key. And it is precisely to avoid this risk that Wright proposes various experiments to the babblers that are ultimately intended to compel them to show that they are indeed a particular instance of sociobiological theory.

But an event came to shed new light on what Wright calls, in this context, anthropomorphism. One day we were facing a nest, he and I, watching the comings and goings of birds helping the parents feed the chicks. The babblers were thus going about their business, whether it be increasing their prestige or responding to the program dictated by the imperious necessity of their genes. Suddenly, during a moment of our observations, we saw a helper land on the edge of the nest and emit the little signal that indicates that he will feed the young. The little beaks stretched out, chirping towards him. But he gave nothing. The chicks were distraught and chirped even more. Did I see this right? Were we in the presence of a cheat? Wright confirmed it: this bird had not fed the chicks. To the question of knowing why the bird had acted in this manner, Wright had an answer. This bird had emitted a stimulus that should have played the role of a variable, following which he (the bird) had verified the intensity of the reaction to this variable that, according to him (Wright this time, but perhaps the bird as well), should allow him to infer the real state of hunger among the chicks. The bird had empirically *controlled* it. This babbler understood the purposes of experimental procedure. This could be put another way: that the behavior of the babbler "tester" conveyed his distrust with respect to what he observed (the chicks *always* pretend to be starving); he needed not only a proof but a measurable proof. To correctly interpret a situation, then, there's nothing like taking control. The babblers won't let anyone deceive them with false stories, and they won't deceive themselves either.

It is unnecessary to insist on the similarity between what Wright offers as interpretations of his observations and the methods that he privileges and that he believes to be the only relevant ones. But if we choose this path, we notice, then, that Zahavi, in a certain manner, proceeds in a similarly consistent way. The life of a babbler consists in constantly observing others and interpreting and predicting their behaviors. A babbler's way of being in the world, in other words, is constantly punctuated by anecdotes—or rather not, for if I put it this way, I borrow the language from the other camp; the social career of the babbler consists in taking a plethora of details important to it and interpreting them. Every bird is compelled into the incessant work of predicting and translating the intentions of others. This is the life of very social beings.

However, these manners, when described in this fashion, prove to correspond just as much to the way that Zahavi himself observes and makes sense of their behaviors: by paying attention to the details that may be important, interpreting intentions, and attributing a complex ensemble of patterns and meanings.

Admittedly, there is nothing that permits us to clarify what initiates this similarity. Did Zahavi build his practice and his interpretations in such a way that they correspond—in the sense of "responding to" in a relevant way—to the way of life of these birds, or has he attributed to the birds the patterns that he privileges within his own practice? This question can also be asked of Wright. Does he attribute to the birds the way that he has learned of "doing science"? Or should one adopt the answer that he himself would give: his way of understanding *corresponds* to the habits that he observes?

Regardless of whether the response to these alternatives is charitable or critical, it can be seen that the meaning of the accusation of anthropomorphism has slid and ties itself to the problem of the relation of scientists to amateurs. It no longer has to do with understanding animals with regard to human motives. It is no longer the human that is at the heart of this affair but rather the practice and, thus, a certain relation to knowledge. The anthropomorphism of Zahavi, of which Wright accuses him, ultimately does not consist of attributing properly human motives to babblers in their resolutions of social problems but instead consists in thinking that the birds use the cognitive procedures of amateurs—collecting anecdotes, interpreting them, making hypotheses in terms of motives and intentions . . .

The question of knowing who's way of knowing and acting is shaped by whom—are the scientists attuned to the ways of the birds, or are the behaviors of the birds shaped by the scientists?—of course remains open. And the answer that might be given for one of the two researchers may not necessarily hold for the other—might it be that one is "well attuned" while the other has "attributed"? I would not say that this is "unimportant," however, because of course it is important, because this changes the ways that we consider not only what "fabricating science" with animals can be but above all what we can learn with them so as to do it in the right way.

G

FOR GENIUS

With whom would extraterrestrials want to negotiate?

The cow is a herbivore who has time to do things. It's Philippe Roucan, a breeder, who suggests this definition. *The cow is a being of knowledge,* writes Michel Ots, for his part. They know, he says, the secret of plants, and they meditate while ruminating: *what they contemplate are the metamorphoses of light from the distant reaches of the cosmos to the texture of matter.* Didn't some breeders tell Jocelyne Porcher that the horns of bulls are what tie them to the power of the cosmos?[1]

I have sometimes thought to myself—and this is surely already the basis for a science fiction novel—that our imagination is so poor, or so egocentric, that if extraterrestrials were to visit the earth, we think it is us that they would want to contact. When I read what the breeders tell us about their cows, I would like to think that it is with them that the extraterrestrials would undertake their first relations. For their relation to time and to meditation, for their horns (these antennae that link them to the cosmos), for what they know and what they transmit, for their sense of order and precedence, for the confidence that they are able to manifest, for their curiosity, for their sense of value and responsibility, or, further, for what a breeder told us and surprised us by: *they go further than us in their reflections.*

If, in neglecting us in favor of cows, this hypothesis on extraterrestrials makes sense to anyone, it would have to be Temple Grandin. It's true that, when she evokes extraterrestrials, it is more often to say that she perceives *us* as such and that she often feels, according to her own terms, like *an anthropologist on Mars.*[2] Temple Grandin is autistic. She is also the most recognized American scientist in the area of livestock breeding. The two are linked. For if she has become an expert, and if she could design the most ingenious facilities and handling systems for animals, and if she can take on the job of her choice with such success, it is, she says, because she can perceive the world just as the cows themselves perceive it.

When she needs to solve a problem in the field, for instance, when the livestock refuse to enter an area where they are frequently steered or when they create problems that generate conflict with the humans who are in charge, Grandin tries to make legible the way that the cows see and interpret the situation. The act of understanding what could have frightened an animal and that we did not perceive, and what elicits his resistance to do what we have asked him to do—to enter into a facility and go through a corridor—allows Grandin to solve problems and conflicts. It may be but a detail, like a bit of colored rag that flaps on a fence, a shadowy spot on the ground that does not appear to us or that may not mean the same thing to us, and the animal is seen to act in an incomprehensible manner.

The fact that she is autistic, Grandin explains, makes her sensitive to environments, a sensitivity that is very similar to that of animals. Her subtle understanding of them, and her ability to adopt their perspectives, in fact rests on something like a wager. Animals, she claims, are exceptional beings, just as she herself is, as an autistic person. "Autism," she writes, "has given me another perspective on animals most professionals don't have, although a lot of regular people do, which is that animals are smarter than we think. . . . People who love animals, and who spend a lot of time with animals, often start to feel intuitively that there's more to animals than meets the eye. They just don't know what it is, or how to describe it."[3] Some autistic people, she explains, are in the mentally retarded range but are capable of doing things normal people are incapable of doing, such as knowing in a fraction of a second the weekday of your birth with just the date or telling you whether your house number is a prime number. Animals are like autistic savants. "Animals have special talents normal people don't, the same way autistic people have special talents normal people don't; and at least some animals have special forms of genius normal people don't, the same way some autistic savants have special forms of genius."[4]

Animals thus possess a remarkable ability to perceive things that humans cannot and a faculty that is just as incredible to remember extremely detailed information that we cannot remember. "I always find it kind of funny," she continues, "that normal people are always saying autistic children 'live in their own little worlds.' When you work with animals for a while you start to realize you can say the same thing about normal people. There's a great big, beautiful world out there that a lot of normal folks are just barely taking in."[5] The genius of animals is thus due to their remarkable capacity to pay attention to details, while we privilege an all-encompassing view because we tend to dissolve these details into a concept through which we perceive. Animals are visual thinkers. We are verbal thinkers.

"The first thing I always do, because you can't solve an animal mystery unless you put yourself in their place—*literally* in their place. You have to go where the animal goes, and do what the animal does."[6] Grandin follows the corridor, enters the barn, traverses the route, follows the path, and watches: the slow turning and rotating of fan blades; shadowy spots along the route that appear like a ravine without bottom; yellow clothing that is frightening because it is too bright, with the contrast "popping into the eyes" like the blinding reflection of light on a metal surface.

It could be thought that Grandin, in describing the procedure that consists of putting herself in the place of animals in order to think, see, and feel like them, refers to what could generally be defined as empathy. But if it is indeed empathy, the term now reveals an oxymoron: what we are dealing with is an empathy without pathos.

It would thus be a technical form of empathy that rests not on the sharing of emotions but rather on the creation of a community of visual sensitivity, on a talent that is much more cognitive than emotional because this is the way that we categorize this type of process. If I'm unable to find the words to account for this event—I, who am inscribed within a tradition in which empathy is taken from the sphere of emotions—I can nevertheless turn to a little experimental marvel, the science fiction novel *Foreigner* by C. J. Cherryh.[7] In a faraway universe, in both time and space, an earthly ambassador is sent to a planet where strange beings, who are apparently very similar to us, live, to create relations with them, speak with them, and attempt to resolve conflicts. Yet, and this is what makes them strange, the beings of this planet know affects, but they are nothing like our own: they have nothing of the interpersonal. There is no love, friendship, hate, or personal affect between them. The entire difficulty of the human ambassador is to understand a relational system that is to this point similar to our own—where people cultivate relations, help each other, kill one another—but whereby he is always tempted to translate these relations into emotional, interpersonal terms. What holds the people, what bonds them together, and what explains their conduct is in fact based on relational allegiances and loyalties that prescribe, like a set of rules, the codes of conduct. And this produces a type of society, and of relations, that is at this point so similar to our own that the hero never stops mistaking the motives and intentions of those who help him or behave like enemies. Though he is mistaken, the transactions nevertheless work, and if the errors that are due to mistakes have consequences, Cherryh ensures that they are not definitively sanctioned. The story is not a moral lesson but about an experimentation that compels the hero to defamiliarize himself from his habits and obliges him to think and to hesitate.

The analogy appeals to analogy, one might say. But the path of science fiction and the example of *Foreigner* invite us to slow down. Are animals "really" like autistic people? Grandin believes so, and with a certainty that is really difficult to go along with for those who are neither animals nor autistic. But the regime of truth that accompanies this affirmation inscribes itself within the regime of the fabulatory wager; it is a pragmatic gesture. In acting *as if* she were dealing with beings who, like her, see the world in a certain way, have a genius for detail and a talent for perception, she happens to get from these beings what constituted the object of the wager: better attunement to the intentions of the breeders and of the animals. And, in fact, there is less violence on farms because of her work. In other words, she teaches American breeders to see and to think about the world with the genius specific to their animals. It is by design that I specify that the breeders are American, for most of the breeders of this country, unlike those I mentioned at the beginning of this chapter, have little contact with their beasts, except on those very specific occasions when they are provided with care and at the moment of their transportation to the abattoir. The type of breeding that Grandin deals with, in other words, only partially overlaps with the meaning that it can have for some of the breeders in Continental Europe, for whom the cohabitation with their beasts, and the act of knowing and loving them, constitutes the essence of their occupation.

Animals are geniuses. Temple Grandin offers a lovely antidote to the thesis of human exception. She has inverted it. It is the animals who are exceptional, just like autistic people are exceptional beings. The analogy, of course, establishes what could pass as equivalences, but she does it according to a system of inversions that problematize these equivalences; the analogy is not immediate but rather rests on the construction of two differences and on putting them together: the difference between humans and animals and between autistic and normal people. Even more interesting, and this is what confers on it the role of antidote, the analogy is based on the retranslation of these differences into *qualifying differences*. The stupidity of beasts [*bêtise des bêtes*] and a human handicap becomes a particular, exceptional, genius talent in being in the world. When developed this way, the comparison reinvents identities. It proposes other modes of accomplishment. It is therefore not comparison but translation. To make the "same" with the "other," sameness with a different otherness. To bifurcate becomings. To construct oneself through stories that make oneself grow. To fabulate.

It is probably not by accident that Temple Grandin remembers—as she recalls her long path and the role her mother played in fighting to spare her daughter, diagnosed as "schizophrenic," the fate of being placed in an institution—the stories that her mother told her when she was a child: that in the night, fairies would sometimes visit houses where an infant had just been born and replace this infant with one of their own. And the humans would find themselves with these little bizarre beings that they could not understand and who seem not to understand them, these children whose minds [*esprit*] disappear in such strange ways and who always remain exiled, these children that the world of our language and of our bonds has such difficulty welcoming, these children who see captivating and frightening things that no one else perceives—in short, children who, like Temple Grandin, introduce invisible and fabulous worlds into our own.

H

FOR HIERARCHIES

Might the dominance
of males be a myth?

A pack of wolves, as stipulated by the Franceloups.fr website when I consulted it at the end of September 2011, "often consists of a dominant couple that has the role as leaders of the group. We call them the Alpha male and the Alpha female. The dominant couple makes all of the decisions for the survival of the pack, its hunting movements, its marking, and territory. The Alpha couple is the only one that reproduces. In the pack the hierarchical order consists next of Betas who come after the Alphas. In the event of a problem for the pack (e.g., the death of the Alphas), the Betas take their place. Next come the Omega wolves, an unenviable position in the pack since the Omegas endure perpetual and daily aggression. In their position of rank, the Omegas are the last to eat the prey killed by the pack."[1]

A description similar enough to this organization can be found in the literature devoted to baboons in the 1960s. The primatologist Alison Jolly, in reviewing the famous conclusions of Irven DeVore's *Primate Behavior* (published in 1965), notes that the discourse of reference for this era, and for the decades that followed, claimed that "the main characteristics of baboon social organization . . . are derived from a complex dominance pattern among adult males that usually ensures stability and comparative peacefulness within the group, maximum protection for mothers and infants, and the highest probability that offspring will be fathered by the most dominant males."[2] It's close enough, with a few similar details; so, for example, among baboon specialists, researchers insist on the role of dominant individuals in the protection of the troop. Jolly, who carried out an overview of the research in 1972, notes that this is the prerogative of the most highly ranked males, and it is even the most clear sign of dominance: "When a savanna-living troop of baboons encounters a big cat, it may retreat in battle formation, females and juveniles first, the big males with their formidable canines last, interposed between the troop and the danger."[3] However, this beautiful model of organization, Jolly concludes, has one exception: the baboons of the Ishasha forest in Uganda, as observed by the primatologist Thelma Rowell, flee in great disorder when they see predators, each according to his own speed. Which is to say that the males are far off in front, and the females, encumbered by their infants, trail behind.[4]

This flagrant lack of heroism—as Rowell herself puts it—was in fact but one of many eccentricities in the behavior of these particular baboons: the Ishasha baboons did not recognize any hierarchy. No male dominated the others, nor did they seem to have the ability to secure privileges associated with rank. On the contrary, a peaceful atmosphere reigned in the troop, aggressive behaviors were rare, and the males appeared to be much more attentive to cooperation than to maintaining the competition that reigns in other groups. The primatologist also reports an observation that is even more puzzling: there did not seem to be any hierarchy between males and females.[5]

This information was welcomed with skepticism from Rowell's colleagues. No baboon has ever behaved this way, so those of Ishasha were an unfortunate exception to the good order that nature has provided the baboons. There had to be an explanation for this. Eventually one is found that should not anger anyone, neither the primatologist, who otherwise "would have observed incorrectly," nor the baboons, who didn't conform to their being baboons—the latter having happened, at the beginning of the 1960s, to the chacma baboons of southern Africa. These baboons paid dearly for their audacity: their observer, Ronald Hall, reported during this period that they were observed to have no hierarchy. They thus found themselves excluded from the species; they were not baboons! A less harsh solution was found for the eccentricities of the Ishasha baboons: their odd behavior must be due to the exceptional ecological conditions from which they have always benefited, namely, the forest, a veritable earthly paradise with its trees that offer shelter from predators, places to sleep, and, above all, an abundance of food. The myth of earthly paradise, and the fall, is never far from the origin myth that the baboons help to reconstruct: the Ishasha baboons remained in the trees and thus did not accomplish the evolutionary leap that their savanna relatives consented to make. All progress has a price, and the latter paid for it by much harsher conditions that brought about intense competition, leading them into an extremely hierarchized organization. This explanation, in ecological terms, may have marginalized the Ishasha baboons, but they nevertheless gave them the chance to remain part of the baboon species and gave credit to the researcher for her observations. With these problems resolved, research could thus continue to accumulate evidence for the universality of hierarchical organization among savanna baboons—and among innumerable other species.

The model had also, at this point, become so inevitable that it determined, in every field, the first point of inquiry. Every inquiry had to begin with the discovery of a hierarchy and the establishment of each individual's rank. And if such a hierarchy didn't seem to appear, the researchers would then invoke a convenient concept to fill in the factual hole: that of a "latent dominance." The dominance must be so well established that it can no longer be seen.

A few years later, at the beginning of the 1970s, Rowell decided no longer to accept the marginalized position to which her baboons had been relegated. Yes, the Ishasha baboons benefit from particular conditions that might account for their deviance. But one must agree on what one calls "conditions": they are not the ecological conditions in the traditional sense of the term but rather the very conditions of observation. In other words, these baboons are exceptions to the model only because they were observed in conditions that did not compel them to obey this very model.

Rowell actually went back over and compared all of the research carried out before her.[6] She was able to classify baboons into two groups. On one hand, one finds animals who are obviously not interested by hierarchy, those for whom it was necessary to invoke the concept of latent dominance, those who were thought to know different selective pressures, like the Ishasha baboons, or those who have been excommunicated from the species, like the chacmas. On the other hand, one finds all of the baboons, in the field just as much as in captivity, who have behaved in a manner expected by the model. Two constants appear. In all of the research in captivity, the baboons are very clearly hierarchized; in nature, dominance emerges in a remarkable way in observational situations wherein the researchers have encouraged the animals to draw it out. A coincidence? Not really.

All of the research in captivity is fashioned after the same model. To study dominance, scientists match up a pair of monkeys and put them in competition for a bit of food, for space, even for the possibility of avoiding an electric shock. The two monkeys are most often complete strangers. In the first test, one of the two will win, for this is the goal of the operation. In the next test, the other will anticipate the predictable result, and if he fights, he will not do so with all of the necessary conviction. Each iteration of the test comes to confirm a more and more reliable prediction, just as much for the experimenter as for the monkeys. Eventually, in the presence of something coveted or a shock to avoid, the one who has lost all hope will step aside and avoid finding himself in the path of the one who became "dominant." This phenomenon is reproduced identically when groups are composed. The lack of space and food inevitably provokes conflicts between monkeys who do not know one another and who are brought together into a social group where the structure is in a certain way determined by the very device of captivity.

In the field, things are no doubt different. Individuals know one another; they are not, in principle, subject to the same constraints. But there is a forgetting of the constraints of research. For if researchers have enticed their baboons with food instead of the practice of habituation, they have most often done so with an insufficient quantity and in only one spot, which thus provokes great battles, after which the dominant baboons are easily identifiable. The researchers have thus reproduced the conditions of captivity in the field. Rowell's verdict will be without concession: hierarchy only appears so well and so stable within conditions where researchers have actively provoked and maintained them.

The model nonetheless continues to permeate research.

Here and there, however, some recalcitrant baboons showed themselves. Those of the young American anthropologist Shirley Strum, known as the "Pumphouse Gang" of Kenya, seemed to want to take up the torch of resistance in the mid-1970s. Strum comes to the conclusion that the dominance of males is a myth.[7] All of her observations are consistent: the most aggressive males, and those classified the highest in the hierarchy if we base this on the outcomes of conflicts, are the ones least often chosen as a companion by the females and have much less access to females in estrus. Against all odds, once a male has an advantage in a conflict, it is the loser that is treated better. He enjoys the attention of receptive females, is given favorite foods, is groomed frequently. The outcome of the conflict, Strum explains, shows that this does not consist of a simple problem of dominance or of access to resources; these notions need to be seriously put back into question to understand the relations involved.[8]

The reception of her propositions will be disastrous. She will be accused of poor observations, even of manipulating her data. "Of course there is a hierarchy among the males of the Pumphouse Gang," she will hear repeated over and over again by the "silverbacks" of universities.

The brutal rejection of her research and the little follow-up given to Rowell's critiques only render more perceptible the difficulty researchers have in abandoning this notion. With Rowell, one can allude to the force of myth in primatology, stemming from a Victorian and romantic naturalist tradition of a dominant male fighting for females, even of a certain form of anthropomorphism, or "academicomorphism": wouldn't relations of hierarchy, after all, characterize the relations between those who write the most about their subject?

One can also think that the reasons for the almost maniacal predilection for this model are tied to the ambitions of a majority of primatologists to confer on their research a scientific basis within a naturalist perspective (☞ **Fabricating Science**). In this respect, hierarchy makes a good object. It confirms the existence of species-specific invariants; it assures the possibility of predictions that are reliable and may be subject to correlations and statistics. But the conception of a society arranged according to the principle of dominance is also taken from a social conception that primatologists borrowed from sociology, according to which society preexists the work of actors (☞ **Corporeal**). This conception, according to Bruno Latour, does not succeed in establishing itself except by obscuring the incessant work of stabilization that is necessary in the act of making a society. The theory of hierarchy would be sort of like a frozen image. There are, of course, many tests of aggression among baboons, and tests through which they attempt to show who is the strongest, but if one wants to construct a relation of order, one cannot do so except by shortening the time of observation to just a few days. Does a hierarchy that fluctuates every three days still merit the name of hierarchy? A hierarchy in which one can claim the conquest of a female is not the same as one that claims a privileged access to food, nor is it the same as one in which the movements of the troop are decided, a role reserved for the eldest females among the baboons. Can it still be *one* hierarchy?

Hierarchy and dominance nevertheless remain quite present in a good part of the literature and continue, for some researchers, to be self-evident. Admittedly, they do concede, "it is more complicated than this." But this does not diminish in the slightest their obstinacy in using them to describe these types of relations (☞ **Necessity**; ☞ *Umwelt*). This is evident in the introductory paragraph on the wolf pack that opened the discussion. This idea of hierarchy still feeds into dog-training manuals, demanding that a master remind his companion, if it tends to forget, who is dominant.

This persistence is all the more surprising because the wolves followed, in this respect, the same path as the baboons. In the 1930s, following the work of the specialist Rudolf Schenkel, the theory of the alpha male is imposed. At the end of the 1960s, the well-known American wolf specialist David Mech will pick it up again and will extend the research in this direction and contribute to popularizing it. By the end of the 1990s, however, Mech will call the whole theory into question. He had followed a pack in Canada for thirteen summers: what had been called a pack was in fact a family, composed of parents and pups who, upon reaching maturity, left their family to form one of their own. There is no relation of dominance, only parents who guide the activities of their pups, teach them to hunt, and teach them to behave themselves well.[9]

The reason for this disparity between theoretical positions is simple and predictable now that we know the story of the baboons: before the thirteen summers of observation, the research of Schenkel and Mech was confined to animal parks and zoos, beginning with packs that were artificially created with individuals who were strangers to one another, confined within spaces in which no escape was possible and with food that was provided by humans. These wolves would try, as best as they could, to organize themselves despite the stress that each of these elements continued to feed. The alphas would thus claim all of the privileges, the betas would compose themselves, and the omegas would try to survive the incessant persecutions. This is the daily spectacle on offer at many animal parks.

And this is the description that continues to prevail in the literature. The theory of dominance therefore seems well and truly destined to persevere as long as humans continue to allow it to exist and put up with it. All of this, one can see, does not pertain exclusively to theoretical problems. Our theories with respect to animals have practical consequences, if only because they modify the consideration that we can have with respect to them. And this goes well beyond a simple consideration, as amply demonstrated by wolves in parks and the answers that are given when one worries about the incessant attacks to which the omega wolves can be victim: "That's just the way wolves are."

The theory of hierarchy has all the allure of an infectious disease in which the virus belongs to a highly resistant strain. Its symptoms, just like its virulence, are readily identifiable and mappable: it produces beings who are determined by rigid rules, beings who are not very interesting, beings who follow routines without asking too many questions. And it contaminates the humans who impose this theory just as much as the animals upon whom it is imposed.

I

FOR IMPAIRED

Are animals reliable models of morality?

During the *Bêtes et Hommes* exhibition, which was held at the Grande Halle de la Villette in Paris in 2007, five rooks, one crow, two lizards, five vultures, a few bustards, and two otters—a brother and sister—were housed between the works, videos, and texts. These animals "in residence" were, according to the wishes of the exhibition curators (myself included), the ambassadors of their conspecifics; as representatives, they posed questions related to the problem of living together and the conflicts that these decisions generate between humans, between humans and animals, and even between the animals themselves (☞ **Justice**). These animals attested to the difficulties related to the fact that they are at present, in an explicit and collective way, implicated in our stories and to the fact that we are today required to explore and negotiate with them the ways they might be interested in this implication.

In making this choice, the exhibition curators knew they were taking the risk of being accused of putting these animals in cages. But they also carefully prepared the modes of legitimation and, above all, ensured that the maintenance conditions for these animals in residence were beyond reproach. It was the otters that took them by surprise.

Everything had started well enough. Day by day, the otters appeared to be acclimatizing to their new environment and even multiplied the signs of well-being. They had, therefore, favorably welcomed the proposals and responded to the expectations of those who had mobilized them. The latter were not expecting, however, that the otters would take the initiative to surpass their expectations. And they certainly hadn't asked that one of the signs of the otters' well-being take the form of an aberration with respect to norms of sexual behavior.

For biologists had already assured them: all scientists today agree that among otters, as with many animals, certain mechanisms impede the attraction of individuals who have been raised together. Clearly the brother and sister otters had decided to provide their contribution to, or more precisely reopen, the old controversy surrounding incest. Furthermore, they seemed to want to prove contemporary ethologists wrong and, by the same token, return to the hypotheses of Sigmund Freud and Claude Lévi-Strauss, who, though surely not specialists in animal matters, had strong ideas on the issue and had made it a criterion of the "properly human"—humans know the taboo of incest, but beasts do not.

Even if the exhibition curators were not worried by this controversy, the fact that their otters contradicted the scientists with such impunity made them fear the worst. It's known in fact that zoos and captive conditions have for a long time had the reputation of "denaturing" their animals; in the domain of sexuality, this accusation generally takes aim at sexual behaviors said to be perverted, in this context definitively known as "unnatural."

Note, however, that a good part of what we know about the sexuality of animals comes from research in captivity. First of all, this may be because its fairly difficult to observe in natural conditions where animals tend to be relatively discrete in these matters, mainly because these activities entail a greater vulnerability. In zoos, however, unless they submit to a disapproving abstinence (which happens often), the animals often have no other choice than to participate in the sexual education of the spectators—and in maintaining biodiversity, we tell ourselves, but that's another problem. Subsequently, sexuality is better known in artificial conditions because that is where it has been studied, which is to say provoked: countless research studies have thus followed or led to the reproductive careers of millions of rats, monkeys, and many others.

Far be it for me, however, to consider that the deviations from the norm that are observed in captive conditions are the unequivocal result of pathological conditions. It is more complicated than this, and generalizations here are of no help. It could in fact be remarked that animals in relatively secure conditions, hardly preoccupied by the presence of predators and the necessities for survival, explore and make visible other modes of relation. Thus the question of pleasure has for a long time been considered irrelevant when speaking of animals. The question was resolved by the double imperative of urgency and reproduction (☞ **Necessity**; ☞ **Queer**). But do animals really have reproduction on their minds? For many of them, clearly, things are quite different. Bonobos have become celebrities in this respect. Among birds, it is now considered that mating might take place for a variety of reasons. The question of pleasure has always been mentioned with great difficulty by scientists, and the speed of most sexual performances only encourages this reticence. Everything changes, of course, if one considers that animals can behave *otherwise* given the opportunity. This sometimes happens. The philosopher and artist Chris Herzfeld, who has spent a lot of time with the orangutans at the Jardin des Plantes in Paris (☞ **Tying Knots**), observed a female engaged in intercourse for close to thirty minutes and who, quite evidently, wanted to prolong this moment. This shows us that animals can deploy another repertoire if the conditions are favorable. Captive conditions are clearly different than natural conditions; but they are not any less real. They constitute in some ways a series of different propositions, and as such, they can be judged favorably, or not, and always *in a certain respect* (☞ **Hierarchies**).

The fact remains that, concerning the two otters, those responsible for the exhibition were rather uncomfortable and imagined the difficulty in using, once journalists caught wind of this affair, and, following them, animal activists and the public, the resources of this argument.

They knew that if they had housed the otters a few decades earlier, no one would have been worried. It would have been perfectly normal for animals, insofar as they are animals, to not respect the rules in force among humans. The prohibition of incest and the control of sexuality have for a long time been among the decisive criteria of human exceptionalism. Noticing their discomfort, however, the biologists who were collaborating in the exhibition reassured the curators. In fact, so they claimed, when animals are in agreeable conditions, this can happen, but hormonal mechanisms would prevent such mischief from having unfortunate consequences. Those responsible trusted the biologists just as they trusted the otters. Animals, however, are not always in tune with their scientists, and as for the trust, it does not always work unilaterally. Indeed, a little while later, the little female otter began to put on weight in a worrying and increasingly significant way. It seems that the hormonal mechanisms did not live up to the expectations of the biologists and organizers. On November 18, 2007, the exhibition website thus announced a happy birth, without clarifying, however, the bond that united the two parents.

It can be seen that in light of this story, what might have otherwise appeared as a natural characteristic was defined, within this context, as the exact opposite: it became seen as unnatural. The fact that it was identified as "unnatural" within the domain of sexuality is not without importance. The otters could have, for example, discovered the use of a nutcracker or started dancing within their enclosure, and this would have been received with enthusiasm, not the disapproval that the exhibition curators feared.

It is worth noting that rarely does this disapproval manifest when it concerns domestic or laboratory animals. Pure strains of rats and mice have been created just by cross-breeding the most closely related so as to reduce the behavioral or physiological variability that might inadvertently contaminate experimental results. For different reasons, the same has been done with breeding livestock animals and with dogs, for whom the value of a pure breed—or that of certain appreciated characteristics— has been held as a guide in matters of selection. The entire process of domestication has been guided by principles that are not necessarily the criteria that animals would apply if they were left free to make their own decisions—far from it.

But in the wild today, with but a few exceptions, endogamy—the act of mating with close relatives—is considered to be generally avoided. The exceptions are for the most part reserved for a few populations that are limited in their options, such as those found on islands. Other exceptions also exist, of course, like the cichlid *Pelvicachromis taeniatus,* the little monogamous and brightly colored fish that lives in the coves and rivers of Cameroon and Nigeria.[1] The females of this species prefer to mate with their brothers and the males with their sisters. Scientists have tried to understand the reasons why these fish transgress a rule that is generally well followed in the animal kingdom, and they think that these fish have in fact been led by natural selection to prefer close relatives to reproduce because the supervision of the eggs and the young, particularly against predators, demands work that is only efficient if the parents cooperate fully. Or, it would seem, the collaboration is of much better quality if the parents know one another. It is always the case that this type of research nicely highlights the recent inversion in ways of thinking. It's the animals who are not respecting the rules of exogamy and who must now provide an explanation. And it had better be well justified!

The sexuality of animals has for a long time fueled the thesis of human exceptionalism ("we're not animals," as it is so aptly put, refers to this dimension of the problem) and has always fed a broad regime of accusations and exclusions—of those who, fittingly, behave like beasts—that runs along a fairly complex dividing line between those that nature tolerates (incest) and those that have been virtuously prohibited (homosexuality). Beasts, therefore, behave like beasts until we change our minds about what behaving like a beast means. Animal sexuality thus always appears as a model, either to follow or to distance oneself from, to enter into culture. This preoccupation still remains relevant, albeit in new forms. The case of the monogamous vole, as studied by the young Swiss researcher Nicholas Stücklin, is exemplary in this regard.[2] The story is all the more interesting because this mole never stopped moving from a commendable attitude of compliance to the model that it was supposed to establish for the scientists, to a pathetic indifference to it.

The prairie vole *Microtus ochrogaster*, with its little ears and yellow belly, is a rodent that lives in the Canadian and American Midwest. In the neurosciences, this vole achieved a certain notoriety thanks to a social behavior that some zoologists attributed around the end of the 1970s: the vole, we are told, is monogamous and biparental, a behavior that is recognized in only 3 percent of the entire class of mammals.

The story as Stücklin tells it begins in 1957, when the zoologist Henry Fitch noticed that while capturing voles for data collection in the prairies of Kansas, it was often the case that a male and female who were discovered in one trap would often be found together in other traps. The monogamy hypothesis that would later be applied was not Fitch's, however. Upon capture, he noted that the female was not in estrus; thus, according to him, the liaison was not of a sexual nature but rather the animals were nest mates who had the habit of venturing out together. If one found itself trapped, the other would try to force its way into the cage to rejoin the trapped one. Furthermore, it was sometimes the case that these pairs consisted of two females. Because Fitch was unable to elicit any sexual activity in the laboratory, he could neither confirm nor revoke the hypothesis of an eventual sexual bond between the "friendly" partners.

But in 1967, some different zoologists resumed observations and focused on a different characteristic to which Fitch had paid little attention: males actively participate in the raising of the infants. Ten years passed and interest in the vole changed: researchers now considered the vole as a candidate for the role of model in the laboratory, understood as a model for human behavior. Monogamy had become a serious issue. Two scientists, H. T. Gier and B. F. Cooksey, focused on paternal behavior as the key to monogamy—generally, when couples are stable, both parents are invested in the care of the young.[3] The male is thus found to be considerate, cooperative, even docile with respect to the one who has become "his" female; he grooms and feeds her, and he even assumes, and *admirably so,* according to the researchers, the role of midwife by attending to the nest and the babies after birth. Only a monogamous being would be devoted in such a manner! The reputation of the vole is established, and researchers will continue to observe the fathers for the next twenty years. The vole, now monogamous, thus begins to interest the neurosciences, in which there is a search for a model of attachment. Laboratory rats saw themselves dethroned; if they provided evidence of maternal attachment, they were completely helpless when it came to being a couple. The vole became the model of the physiology of love (that is, for humans) and of the formation of couples (that is, heterosexual couples). Neuroendocrinological research will see new growth. The mammalogist Lowell Getz and the behaviorist Sue Carter consider yet another possible fate for the monogamous vole. If they can demonstrate the chemistry of relations, they must therefore also be able to establish a model of pathologies of these same relations among humans, and thus store a considerable stock of many different symptoms of social dysfunction. On the condition, of course, that the vole remains monogamous . . .

However, it seems that the model proved to be less perfect than it first appeared to be. Researchers at first discovered the existence of "vagabond voles." For a period of their lives, a nonnegligible selection of voles that were supposedly faithful and monogamous were actually traveling around and frequenting with other voles. Then, some DNA studies came to confirm what had already begun to be suspected: the vole was unfaithful.

According to the research, 23 percent to 56 percent of babies came from "extramarital" relations. And these deserving fathers were in fact occupying themselves with the descendants of another, which, from the point of view of the rules of selection, is not recommended.

This news is certainly embarrassing and, as Stücklin rightly underlines, compromises in an unfortunate way the investment in the vole as a model of the human couple.

And yet . . . one begins to wonder. The vole may not be monogamous, but what does it mean to be monogamous? And are humans, after all? Do they form relations as long lasting as that? Do they share in the raising of their own kids? It's not far off from the story of Freud's kettle—I never borrowed your kettle, and besides, I returned it in good condition, and it was already broken in the first place.[4]

The notion of monogamy will thus undergo a serious expansion. Sexual fidelity will be distinguished from social attachment. The monogamy of the vole, now a social monogamy, will remain intact. The problem is all the more resolved because it is the attachments and pathologies that result from its inhibition that are of interest in the research on the neural bases of human behavior.

In the laboratory, however, this great diversity risks compromising the reliability that can be attached to the reproducibility of behavior. If the vole is a bit fanciful in nature, its monogamy in captivity is the result of constraints imposed by the laboratory, and it would thus be an artifact. In this respect, rats prove to be more predictable and reliable—researchers have also worked hard to reduce this variability to the greatest extent possible by imposing, most notably, sexual choices. However, because rats do not seem to form attachments, we cannot count on them.

Researchers will therefore modify the definition of what interests them: what is common to both voles and humans? The variability of their behaviors, of course! The vole can still continue to be the model par excellence.

We should rejoice. Do we not prefer a world that is marked by diversity? Isn't such a world more interesting? Doesn't it promise more curiosity, more attention, more hypotheses? Without hesitation, I'd say (☞ **Queer**). But I believe that the voles ask us to hesitate. For variety is in the process of becoming a moral response, an abstract and all-terrain response, which indicates that we are going too fast and simply making variety into a generality.[5] In other words, variety is becoming a response rather than constituting a problem.

This will not be seen if one approaches the affair of the voles in the schemas that have become common for deconstructing this type of story. In fact, the modifications of "interesting" vole behaviors can be understood quite well, in the sense of a greater diversity of ways of organizing conjugality, like constituting a faithful trace of the evolution of our ways of organizing ourselves. This could have already been suspected when researchers announced, during a period that coincided with the emergence of feminist movements that called into question the traditional distribution of tasks around children, that voles are excellent fathers to the family, in the new sense of the term—they not only have to earn the dough but also have to bake it. The practical conditions that are associated with these new habits should, however, not be neglected: the vole who obstinately refused to reproduce in captive conditions with Fitch in the end accepted to do with scientists who came along after.

If one takes an interest in more recent research, it's also true that the discovery of the variability of conjugal practices recognized among voles looks very much like the innovations in contemporary Western practices—not forgetting that, from the outset, it's for the inhabitants of this part of the world that the voles must provide evidence. Would it be enough to take note of this variety of habits and to legitimize different forms of couples and different definitions of the family? And to make this variety a sign of natural variability?

It is worth considering. But Stücklin proposes another hypothesis, and this one invites us to slow down a bit. One must, he says, take note of the program modifications and the research agendas that this new vole provokes.

The vole, do not forget, is mainly requisitioned by questions related to the psychopathology of bonds. Attachments, in this respect, can undergo testing in a number of experiments that will show how one can prompt their failure or inhibit them and will measure the consequences of these tests according to a model of failure: in other words, to create situations "without attachments" or situations of disturbed, traumatized, inhibited

attachments . . . whose results mimic mental troubles and social pathologies (☞ **Separations**; ☞ **Necessity**). The more the attachments vary, the
more exploratory paths can be opened, and the more pathological conditions can be considered. In other words, if I allow myself to be guided
by the way in which Isabelle Stengers invites us to pay attention to the
transformations imposed by "making science," the "variety" exhibited
by the vole is translated into the regime of possibilities of "variations":
which can, because it varies, become an "object of variation," that is to
say, in this context, a variable to manipulate.

Here one may fear for the vole. His depravities and infidelities could
relieve him of the task of transforming a model of social conformity into
a natural model; the inventiveness of his ways of being faithful, or not,
is worth him being newly involved in our stories. Not that these stories,
I'm afraid, have much chance of interesting him.

J

FOR JUSTICE

Can animals compromise?

One of the park rangers at Virunga National Park, himself originally from the Lega tribe of the east central part of the Democratic Republic of the Congo, once reported to my colleague Jean Mukaz Tshizoz that, in some villages, an agreement was reached between lions and the villagers. This agreement, Tshizoz tells me, was not unknown to him, for his grandmother had spoken to him about it; one can find instances that are quite similar among the Lega of other regions, among the Lunda of Katanga, and among other Bantu peoples. According to this contract, peace reigns between the villagers and the lions so long as the latter leave the children alone. But if a lion attacks a child, a retaliatory response is organized as quickly as possible. The villagers set off in pursuit of the culprit by playing a specific song with their tam-tams, a song that is designed to warn the lions that a hunt has been organized to punish the act. When they encounter a solitary lion, generally the first lion they come across, they kill it. The criminal is punished. Of course, one can wonder if it is the real culprit that has been punished due to it being the "first one seen." It seems that the answer to this question is affirmative. On one hand, the villagers explain that if a lion is by itself, far from its pride, there is a good chance that the animal is in fact a desocialized individual and that it is thus this desocialization that explains the brutal transgression of codes. On the other hand, they say that the culprit is never far away, which is decisive proof of his culpability, because his proximity indicates that he has developed a taste for human blood. Additionally, it is a sign that the lion will forever be a deviant. The punishment will prove to be doubly pertinent, as a measure that is both punitive and preventative. Once the culprit has been punished, such an incident should not happen again, especially because, as Tshizoz explains, the playing of the tam-tams has the explicit goal of, in his words, "making an impression," and the animals will have understood this. Another time, another place. In spring 1457, a horrible crime disturbed the people of the village Savigny-sur-Étang. The body of a five-year-old boy was discovered, murdered and half devoured. The crime had some witnesses who reported the suspects. The latter, a mother and her six children, were brought before the authorities. They were pigs. Their culpability left little doubt, for they were discovered to have on them traces of blood from the murdered boy. The guilty pigs found themselves in court, before a packed room. Their poverty earned them the right to a court-appointed lawyer. The evidence was examined and, in the face of the facts, the debate hinged on legal issues. In the end, the mother was condemned to be hanged. The verdict for her children, conversely, benefited from the convincing argument of their lawyer: they did not have the mental capacities that could, in the eyes of the law, prove them responsible for the accused crime. They were thus placed in the custody of the state so that their needs were looked after.

These two stories don't have much in common, of course, if the point is just that humans and animals find themselves dealing with conflicts according to rules that arise from the areas of justice.[1] The differences could be emphasized, for they are important and numerous, but what interests me more than the differences are what these modes of conflict resolution assume: that the animal is the author of his actions and can be held to respond. As evidence, in the case of the lions just as much as with the pigs, it wasn't just anyone who was punished, in any which way. It was this particular lion that transgressed and not an other; it was the mother that could be held accountable, not her children.

Many animals, in both Europe and the colonized Americas, have been prosecuted by the book, and one can trace these trials right back to the start of the eighteenth century. The Church undertook prosecution when animals destroyed crops, when they were implicated in sexual relations with humans, or simply when they were believed to be possessed or involved in sorcery.[2] The secular courts, for their part, took charge of cases of personal injury to others.

These practices now seem exotic, irrational, and anthropomorphic, and they are often subject to mocking disbelief. These trials nevertheless testify to a wisdom that we are relearning to cultivate here and there: that the death of the animal may not be self-evident. The court had to intervene to make a decision, with all of the slowness and all of the problematizing that comes with the very forms of the judicial dispositive. In addition, trials relating to the damage of crops or human goods by animals frequently involved a search for a compromise. Evidence of this can be found in a judgment made in 1713, in the region of Maranhao, Brazil. Some termites were found responsible for the destruction of part of a monastery that collapsed due to their activity in its foundation. The lawyer assigned to them pleaded in a clever way. Termites, he said, are industrious creatures: they work hard and have acquired from God the right to feed themselves. The lawyer even called into doubt the culpability of the beasts: according to him, the destruction was nothing but the sad result of the negligence of the monks. In light of the facts and arguments, the judge decided to require the monks to provide a woodpile for the termites; and as for the termites, they were ordered to leave the monastery and limit their commendable industry to the woodpile.[3]

These compromises resemble, in some of their aspects, those that we are in the midst of reinventing with animals.[4] They are evident in the case of protected species with whom we must learn to compose ourselves, whether it be vultures that arrive in great numbers to the offer that is extended to them in the form of livestock burials, wolves with whom cohabitation is not without its problems, otters, marmots . . . The responses of these animals to our protective propositions convey an "excess of achievements," and we must now imagine solutions, always cobbled together, to deal with the consequences of these excesses of achievements. How can the vultures be convinced to leave room for other species? How can we negotiate with marmots who are having fun in the fields that the farmers want to farm? No longer using dead stock burial pits, and asking vultures to manage things for themselves, might diminish the appeal of these sites but lead to other consequences that one must learn to face: some vultures might renounce their scavenging practices by attacking lambs instead. The farmers must therefore be negotiated with. As for the marmots, there was a period when some volunteers were brought together to capture and move them. Over the years, however, the volunteers became more rare and less available. Contraceptive solutions were thus considered, but this too raises problems, notably with ecologists, who protest against such unnatural solutions. But this is precisely what compromises are, as the philosopher Émilie Hache has analyzed so well. It is not a question of compromising morals, as has been pejoratively thought for so long, but instead our principles insofar as they are too narrow to "correctly take into account." "What matters," she writes, "to those who make compromises has less to do with judging the world in light of principles than to treat properly the different players with whom one cohabits; and, for this reason, to be prepared to make arrangements with the latter."[5]

These new ways of compromising have for some time contaminated relations with other species who, though not benefiting from protective laws, prove to have elicited relatively similar consideration. A few years ago, some crows took up residence in a large abandoned garden in a suburb of Lyon. The cohabitation became more and more difficult: the crows were too many, too loud, and their feces proved to be an unbearable nuisance. Complaints from residents multiplied across the municipality, which in turn decided to send in hunters. The neighborhood then protested this: no one wanted these crows to be killed. A solution was then found. Just after the crows' egg-laying season, falconers arrived with their hawks and falcons with the mission to convince the crows to nest elsewhere—ensuring that the brood did not hatch seemed to be the most decisive argument in this matter. No one can say that this solution was just or ideal, and I can't help but recall my obvious discomfort in hearing the despairing cries of the crows who were panicked, abandoning their eggs to escape the attacks. All that we could hope for was that the crows would find a place to relocate where cohabitation would be less problematic; this could not, however, be guaranteed. This solution had nothing innocent about it: we were not innocent, and we did not expect the crows to be either. We learned the difficult art of compromise.

Returning to the practices of trials from which some analogies can be drawn, the latter, however, have a character that remains strange for us: it is not just that the beasts are defended by lawyers, which in a certain way gives them the status of personhood, but above all that they are credited with rationality, will, motives, and, in particular, moral intentionality. Putting them on trial, in other words, adheres to the idea that animals can have a sense of justice.

This idea has not entirely disappeared, but for a long time it has been confined to what one calls "anecdotes," a term that simultaneously denies any importance and any reliability to the observed events (☞ **Fabricating Science**), which is to say to the testimonies of breeders, dog owners, zookeepers, or trainers. Anecdotes are coming back with greater frequency and a renewed vigor in essays today that plead in favor of the better treatment of animals, even their liberation. Animals that run away, revolt, or are aggressive toward humans act deliberately, as attested by their rebellion of conscience to the injustice of which they are victims (☞ **Delinquents**). With respect to scientists, the idea has been slow to catch on. There are many reasons for this reluctance. I would simply point out that in 2000, the psychologist Irwin Bernstein recalled, to some colleagues who were probably in the process of going astray, that morality among animals appears doomed to remain outside of the domain of measurement techniques available to the sciences.

If an idea similar in meaning to justice—or injustice—only begins to appear within studies around 1964, and even here in a manner that is still relatively cautious, I nevertheless discovered the intuition in an experiment carried out in the early 1940s by the biologist Leo Crespi.[6] Admittedly, he is not speaking about justice or injustice, but these notions are not far off. In the beginning, Crespi explains, his research in fact focused on the propensity of white rats to indulge in games of chance—which, according to him, earned him the reputation of promoting roulette and vice among the rodents. Since the results were not very convincing, Crespi decided to interest himself in another problem that seemed to emerge from his research, namely, the effect of variation of encouragement given to the rats—what are traditionally called reinforcements but what Crespi calls "incentives." He observed that while the rats were running in a maze, they would maintain a steady, average speed so long as they obtained the expected reward. But if, once the results have stabilized, the reward is increased after one of the trials, one finds the rats running much more quickly in their next trial, and they even run more quickly than those who had received, since their first trial, the augmented amount of food. It is therefore the contrast that is important, namely, the difference between what the rat feels it is owed and what it actually receives, and not the amount of incentive. The opposite effect is also observable: if the reward is diminished in the course of the procedure, the rats will slow down considerably in the following test. Crespi believed that the rats manifested, in the first case, what he called "elation" and, in the second, a reaction of deception—in some writings, he at times speaks of "frustration," at other times of "depression." I suspect that the choice of the latter term is not without relation to his more promising potential for studies in human pathologies (☞ **Impaired**). To be sure, this research has not resulted in the audacious proposition according to which the "disappointed" rats had the feeling that "it was not just," but the fact that Crespi's experiment is mentioned frequently today in the work on animal welfare testifies to its speculative potential: the animals could "judge" the situations that were proposed to them.

In 1964, Jules Masserman and his colleagues will meanwhile show that when rhesus monkeys are given the choice between "secur[ing] food at the expense of electroshock to a conspecific" or abstaining from eating, they choose abstention.[7] In this experiment, the rhesus monkeys are placed alone in a cage with two compartments separated by a one-way mirror. In the first phase, one side of the cage is occupied by a monkey who the researchers train to pull a chain when a red light illuminates and another chain when there is a blue light, which leads to the arrival of some food. For the next test, the researchers place a conspecific in the other compartment. The one-way mirror is oriented in such a way that the conspecific can be seen by the first monkey in his compartment. One of the two chains, from this moment on, always delivers food but also administers, at the same time, an electric shock to the conspecific who is on the other side of the glass. Thanks to the apparatus, the monkey who works the chains can see the consequences of his actions on his companion. The results are clear: a large majority of the monkeys avoid, from this moment on, touching the chain that delivers the shocks. Some even go so far as to opt for total abstention and receive no food at all. The monkeys prefer to suffer hunger than to inflict pain on their companions. To be sure, the conclusions of the researchers still do not raise questions of justice or fairness; they advance, with a prudent use of quotation marks, the possibility of "altruistic" conduct, and this time without the quotation marks, they speak of protective behavior, noting that the latter is observable in several other species, and thus suggest pursuing research in this direction. This suggestion was heard; other animals, such as rats, were invited to the test. They agreed with Masserman.

Quite recently, however, the idea that animals could explicitly have a sense of justice and injustice has emerged within laboratory research. Owing to the favorable revival of interest that research on cooperation has recently experienced and encouraged, this idea has provoked a few studies. In 2003, psychologist Sarah Brosnan published an experiment in *Nature* that would become famous.[8] She subjected a group of capuchin monkeys to a test intended to evaluate their sense of justice. The group was composed of females, to whom the experimenters suggested the exchange of cucumber slices for some tokens that had been offered earlier. This type of test fits into the general register of those that are said to test "cooperation," where exchange is considered a cooperative act. The choice of females for the experiment is, for its part, justified due to the characteristics of the social organization of capuchins: in the wild, females live in groups and share food, whereas males are more solitary. The exchanges develop without difficulty in the normal conditions of the experiment, wherein the capuchins seem eager to cooperate—and likely think the same with respect to their researchers. But if one of the capuchins witnesses a transaction in which one of her peers receives a grape, which is much more attractive, instead of a slice of cucumber, she will refuse to cooperate. This withdrawal is exacerbated if the partner receives a grape without having offered anything in exchange—"without any effort," the researchers say. Some capuchins thus refuse the cucumber and turn their backs on the experimenter, whereas others, on the other hand, accept the cucumber . . . to throw it back in the researcher's face. The researchers concluded that the monkeys could judge the situations and could characterize them as fair or not, and that cooperation probably evolved, for some species, on the basis of this possibility.

The proposition could be addressed to other animals, but things turn out to be more difficult for them. Monkeys have benefited for a long time from a "hierarchical scandal," as raised by the primatologist Thelma Rowell: researchers give them much more credit because they are our close relatives.[9] And the more we credit them with sophisticated social and cognitive competences, and experiment with them, the more they seem to merit the credit bestowed, and thus the more the researchers are encouraged to pose different questions that are even more complex. Other animals that are considered more primitive, less intelligent, and less talented have often not been entitled to such consideration from scientists—though things are gradually changing for many of them, which has earned them the nice title of "honorary primate" (☞ **Pretenders**; ☞ **Magpies**).

Marc Bekoff, the biologist who specializes in cognitive ethology, is conscious of the difficulty encountered by animals who, to be seen as morally or socially well-equipped, belong to neither a great ape species nor the privileged few who have acquired the title "honorary primate." How can one show, in a way that will be received from a scientific point of view, that animals conduct themselves in ways that are "just," that they implement a whole repertoire of social manners, and that they know very well how to judge what is involved in acting in an "unjust" way? There is nothing obvious about morality, and it resists the burden of proof. However, Bekoff says, this is not the case with play. It is easy to recognize when an animal is playing, and when one carefully observes animals playing, it is clear that play implements a well-defined sense, on the part of the animals themselves, of what is and what is not just, of what is acceptable and what is subject to disapproval—in short, the manners and codes of morality.

When animals play, they employ a range of behaviors that are relevant to their other spheres of activity: they attack, play dead, roll on the ground, lie down, wrestle, follow one another, growl, threaten, run away. They are the same gestures that are found in predation, aggression, or fighting, but they have changed meaning. If misunderstandings are rare, it is because playing can only exist on the basis of an agreement and attunement [*un accord*][10] that is constantly being expressed and actualized: "right now, this is just playing." This attunement is what gives meaning and existence to playing. The gestures are the same ones as those they have appropriated, and yet they are different, for they are constantly accompanied by a code of translation—and with multiple exchanged looks to see that the translation has been effective in connoting the system of action.

Play, Bekoff highlights, falls within the domain of trust, fairness, and cooperation.[11] Trust flows, in particular, from the fact that playtime is safely marked, a time during which transgressions and errors are pardoned and excuses are easily accepted, and that play follows rules but is not defined by them. Fairness comes out of the fact that, within the rules of play, no animal profits from the weakness of another, unless it is being put to the service of play itself. Cooperation has the same condition: no animal plays against his will, and no animal plays with another animal that does not want to play, except through a misunderstanding that is quickly resolved. This is a risk, and it is never absent. Play enacts principles of justice, and animals are able to differentiate between those who are in tune with these principles and those who aren't, for better or worse. An animal who is unable to control his strength or cannot adopt new roles, who cheats, who diverges without warning from play to real life, who is aggressive, who, in short, does not abide by *fair play*, will find himself without any partners to play with after a few of these experiences.[12]

But play is not simply the enactment of codes of conduct. It requires something more that is not explicit within the form of codes, which is difficult to put into words but which is clearly recognizable when two animals are playing. There is, Bekoff says, a "spirit of play." It is play. It is joy.

Play only exists so as to construct and prolong this "spirit of play." This spirit, or humor, is what allows play to be play and is what gives play its context for translation; it is what realizes and creates the attunement [*l'accord*] between partners. The spirit of play creates this attunement, but it is also at the same time created by it; it consists more of an *attunement* [accordage] that designates this event through which rhythms, affects, and flows of vitality are created and attuned.

"This is still play no matter what I am going to do to you": the distinct quality of their actions, their "mood," and the constant looks that are exchanged are acts that both "say" what is in the act of happening (children's "pretend that . . .") and "do" so that it happens and continues to happen (playing again). In other words, when animals say what they are doing, they are doing what they say. The basis for a trusting relation cannot be more clearly defined.

Even if the former terms are not Bekoff's own, I have little doubt as to whether he would adhere to them. Play has for a long time been subject to functionalist interpretations—it could have value as preparation for actions later in life, it allows young animals to initiate conflicts tied to hierarchy, and so on—but with Bekoff, it is a privileged moment to learn what is done and what is unacceptable, to learn how to conduct oneself in a "just" manner, in terms of what is expected, and to judge the ways that others respond to this pragmatic ideal of "rightness" [*justesse*]. Play builds possibilities of trust. Animals learn to "pay attention," for otherwise "it is no longer play." They learn different roles, different possible modes of being, such as that of pretending to be small when one is big, weak when one is strong, that one is playing with a partner who is younger or more fragile, angry when one is happy, and they learn all of this *relative* to an other. Play deploys and cultivates multiple modalities of being granted to others, according to codes of what is just [*juste*] and with the grace of joy. It is to say, then—if I can take up Haraway's proposition and extend it to Bekoff's research—that animals learn, in play, to be responsible, that is, to respond.[13] They learn to respect, to hold in regard, as the etymology suggests. This is what animals do. Concretely. Morality is incredibly funny *and* serious, profoundly joyous *and* grave. This can be learned, among the animals, in laughing with an animal's laugh.

To be sure, the significance given to these terms—*just, attunement, response, respect*—goes well beyond their scientific acceptability. Bekoff encountered this over his entire career through numerous debates and controversies with his colleagues. How many times has he heard "this isn't scientific"? That these terms go beyond what is scientifically accept-able, in terms of play, is ultimately not all that surprising. For if there is something that play does, it is precisely that it plays with meaning and breaks from the literal. Play is the paradise of homonymy: an action that, in one context, translates fear, aggression, or a relation of forces, in another context rearranges, unmakes, and remakes itself; play no longer signifies what it seems to signify. Play is the site of invention and creativity, the site of a metamorphosis of same into other, just as much for beings as for meanings. It is the very site of the unpredictable, but always according to rules that conduct this creativity and its *adjustments*. In short, justice with the grace of joy.[14]

K

FOR KILLABLE

Are any species killable?

Two billion three hundred eighty-nine million pounds of farm animals died in 2009. They were eaten. If one wanted to evaluate the total weight of dead animals, one would have to add to this number those that were killed due to hunting, car accidents, old age or disease, euthanasia, eating by nonhuman predators, and elimination for sanitation reasons or those that were otherwise culled because they were no longer productive. I'm certainly forgetting many more.[1]

How many pounds of humans disappeared over the course of the same year? One does not ask these kinds of questions, or, to be more precise, they are not asked in this manner. If the question of the number of deaths among humans is raised, it will either be calculated according to an average, or be inferred, or be part of a distribution. Under no circumstances will it be measured in pounds or tons but rather by "persons": twenty-five thousand people die every day from malnutrition, eight thousand from AIDS, sixty-three hundred from workplace accidents. I could lengthen this list and find, without much difficulty, the numbers for road accidents, violent deaths, deaths due to drugs . . .[2]

The distribution of these numbers and how they are collected indicate something about our relation to these deaths: these numbers do not convey mere information, and they are not limited to a statistical presentation of the world. Behind all of the work that goes into collecting and modifying these numbers, there is the mark of a *cause,* not only in the sense of causality but above all in the sense given to the term by Luc Boltanski and Laurent Thévenot.[3]

According to these two sociologists, a cause results from the collective work of production of an identity that aims to mobilize, in order to denounce and stop an injustice. Whether these deaths are due, therefore, to AIDS, to workplace accidents, or to malnutrition, there is an equivalence at the heart of each of these categories: they are all unjust because they are avoidable. What ties them together is that they would not have happened if something had been done, if those who have been victims were taken into account, if their causes had been acted on behalf of, whether through prevention programs, a different distribution of wealth, a different organization of work...[4]

For Boltanski and Thévenot, the constitution of a particular "cause" requires "desingularizing" [désingulariser] the victims: it is only through their deaths that victims are presently defined. The denunciation of the two billion three hundred eighty-nine million pounds of dead farm animals that can be found on websites is a part of this operation. The semantics can obviously be a little bit different from what I've just used, for example, 2.389 million tons of meat were consumed over the course of a year. Pounds and tons do not die; they are consumed. There is not just one cause, or many causes, behind these numbers but consequences waiting to be mobilized: the environment, the fate of developing countries, the ozone layer, the health of meat eaters, and the beasts themselves. Dead animals weigh on us, but the weight is distributed: in the methane emissions of cows, in cardiovascular diseases for meat eaters, in the tons of grain that are fed to beasts, in trees that are cut down as part of the deforestation that is required to grow theses tons of grain...

Furthermore, one can see that the desingularization does not operate in a consistent manner: animals that are killed are translated into pounds of meat, deceased humans into persons. It's true that it is the logic of consumption, and the logic of denunciation, that guides the meaning of the translation, at least for the animals. This translation allows, and is particularly aimed at, all those who might be concerned by the effects of intensive animal farming, whether or not they are concerned with the plight of beasts, to rally to the cause: if you are not sensitive to the plight of animals, perhaps you are to the consequences of deforestation that arise due to the agriculture that is required to feed them; and if you don't care about deforestation, perhaps you care about the effects of methane on the ozone layer; and if you are among the climate skeptics, perhaps the question of your own health will succeed in moving you.

But one may wonder about the pragmatic effect of this line of argumentation. The desingularization that creates the "cause" and that is at work by means of the measurable unit of weight not only seems to be a fairly dangerous weapon to manipulate—even according to activists who have realized that the facts invoked lend themselves to dangerous discussions that render them vulnerable—but in a certain manner perpetuates what one could call the effects of the ontological rupture: humans and animals are at this point so ontologically different that their deaths have no possibility of even being thought together. When dead, humans are bodies, corpses; animals are carcasses or cadavers, if they are not destined for consumption. The human cadaver exists, of course, but this designation covers specific situations. In the most general terms, if I go by the news or crime novels, for example, cadavers designate transitory states still "awaiting" resolution. One speaks of a cadaver when a dead body is discovered and has not yet been—or cannot be—"reappropriated" [*approprié*] by those who knew it as a person. The cadaver will remain a cadaver only until it is "reappropriated," at which time it is given back to friends and relatives and the cadaver becomes the "body" of the "deceased": it is a death "for others," a death which thus begins its existence as a death under the protection of the living.[5]

To speak in terms of tons or pounds is to participate in what Noëlie Vialles describes as an operation of translating, materially and semantically, the eater of animal flesh into a "sarcophagus."[6] In her analysis, this expression designates the increasingly pronounced tendency to efface everything that might remind us of the living animal, all that "reminds us too clearly of the animal, of her form, her singular life, and her killing." The concealment of killing is obvious today, as abattoirs have all disappeared from town centers. All that reminds us of the living animal, the animal as a living being, has also disappeared. The most recognizable features of what animals once were are now hidden. Anyone born before 1970 can attest to the fact that there has been a gradual withdrawal of enthroned calf heads occupying merchants' stalls and of the bodies, still whole and sometimes unplucked, of chickens or game birds. The culmination of this concealment can today be found in the hamburger, which constitutes nearly half of the consumption of beef in the United States.[7]

This transformation of the dead into something else that no longer recalls its origin results from the work of what the sociologist Catherine Rémy calls the "deanimalization" of the animal.[8] The process operates inversely to what I have just described with respect to humans: in the abattoir, the animal passes from the state of a body to that of carcass. Practices of consumption will guide the metamorphoses that follow. We now speak, as Vialles emphasizes, of *pork, beef, veal*. Parts of the animal body are translated into modes of cooking: the roasted joint, the parts for stewing, the piece for braising. The dissimulation also operates materially through the methods of cutting, which results in new terms that are for the most part unrelated to anatomy: the prime cut, spider steak, leg, ham, rib steak, rack, side, filet, sparerib, shank, cutlet. The pieces of meat derive from a process that Rémy calls "disassemblage," as if the new order that was assigned to them—prime cut, cutlet, filet—were the natural order. As if it was obvious and not at all problematic. The process of disassemblage that is employed in the abattoir is presented, by this configuration, as though it were the result of a "smooth" transformation. It is especially "smooth" because it effaces, materially and in the imagination, the violence that preceded this transformation. One might remember the image of Tintin in America, when he is amazed to witness, at an abattoir, the transformation of a standing cow into corned beef, sausages, and cooking fat.[9]

Through the consumption of the animal, writes Vialles, one is looking for "effects of life," but one wants them "separate from the living being that gave them substance." In short, our practices are practices of forgetting.

On what grounds should practices of ignorance be reproached? If knowing only aims to modify our relations to ourselves without changing anything about our relations to the world we inhabit, the denunciation is pointless. It only has meaning if it obliges us to think, hesitate, and slow down. It is in this respect that talking about meat consumption in terms of tonnage seems problematic. If, strategically, this maneuver unites a wide variety of interests, creates a "common cause," and inspires a decline in meat consumption, and thus a decrease in industrialized farm-raised animals—a notorious oxymoron—it signals at the same time a problematic proximity to the way that these animals have, more precisely, become no longer *farmed-raised* animals but *products*, like goods of consumption. Thus speaking of the death of animals in these terms comes dangerously close to the reproachful language used within the very practices that contribute to the desubjectification of the animal, practices that are called—the designation is eloquent—systems of animal production. The way that we condemn what we eat uses, and thus ratifies, how we produce what we eat. A simple glance at the pork industry will suffice: there are numbers upon numbers, tons and percentages, comparative graphs and colorful circles that visually reproduce the distribution of the same numbers that are more familiarly called, in applied statistics, pie charts. As emphasized by the sociologist Jocelyne Porcher, the performance in the systems of production is what gives meaning to this work. She states that since the 1970s, when the rationalization of production was established, "the pork industry has collected an incredible amount of data that is meant to account for the work accomplished."[10] She adds, "The production of data in the end takes the place of thought."

Data eventually has a role similar to that of the logic of the sarcophagus: to prevent thinking, to forget.

Donna Haraway notes that, statistically, the most frequent form of human–animal relation is the act of killing. Those who might doubt this have probably forgotten all of the massacres of the last few years, whether due to mad cow disease, avian influenza, foot and mouth disease, or scrapie. To not take these events seriously, she says, is to not be a serious—and responsible—person in the world. Knowing how to take it seriously, she adds, is far from obvious. Regardless of the distance that we are tempted to maintain in relation to these events, "there is no way of living that is not also a way of someone, not just something, else dying differentially."[11] For someone and not for something: it is not the act of killing that has led us to exterminations but the fact of *making beings killable*. Of course, she also says, ethical veganism takes note of an unavoidable truth, that of the extreme brutality that has become normal in our relations with animals; however, a "multispecied" world requires, to have any chance of existing, simultaneously contradictory truths, those that emerge if we take seriously not the rule on which human exceptionalism is based—"thou shalt not kill"—but rather another rule, a rule that makes us face the fact that eating and killing are an unavoidable fact of the relations that tie together all mortal companion species: "thou shalt not make killable."[12]

What we have to find, Haraway continues, and find outside of the eternal logic of sacrifice, is a way to honor. And to honor in all of the places where "companion species" live, suffer, work, die, and eat, from the laboratories that bring humans and animals together, to the domains of animal husbandry, all the way up to our table.

This way of honoring still remains to be invented. This invention requires that we pay attention to words, to the ways of saying that validate ways of acting and being; it requires us to hesitate, to invent new tropes—to trip, as etymology reminds us[13]—to cultivate homonymies that remind us that nothing is obvious, that "nothing goes without saying" (☞ **Versions**).

In this respect, I like Porcher's proposition. She suggests that the animal that has been killed, whether for eating or other reasons, and that who we need to learn to account for in a responsible way is a *deceased* [*un défunt*]. A deceased, not a carcass, pounds, or food product: a being whose existence continues in a different mode among the living and for whom he provides food and sustenance. A deceased whose existence persists, if not in our memories, then in our bodies. It remains to be learned how to remember, to learn to "inherit in the flesh," as Haraway proposes, to learn to make history together with companion species whose existences are so entangled, one with respect to the other, that they live and die otherwise.

Cary Wolfe extends Porcher's proposition when he takes up the question that Judith Butler posed following the tragedy of September 11, 2001: "Whose lives count as lives?"[14] The question as to whose lives are important, or are claimed to be important, results from another more concrete one: "what makes for a grievable life?" Admittedly, Wolfe notes, Butler excludes animals from among the lives that grief claims through a loss, so he takes it upon himself not to betray them by extending the question to animals. For Butler, he says, this demand imposes itself on us because we live within a world in which beings are dependent on one another and, above all, are *vulnerable to* and *for* others. The question of vulnerability, however, does not return the animal to a status of passive or sacrificial victim. And this is the difficulty, it seems to me, of which Wolfe steers clear when he resituates these lives claiming to bear grief in the concrete and everyday dimensions of interspecific relations, dimensions that create a particular form of "common vulnerability" evoked by Butler: "Why shouldn't *non-human* lives count as 'grievable lives,' particularly since many millions of people grieve very deeply for their lost animal companion?"[15] This question is not posed to remind us of the banality of an experience; it is, I believe, essential to Wolfe's approach. For, posed in this way, vulnerability does not align itself according to the status of the victim, and it is not a simple identification of fragility. This vulnerability emerges from active involvement in a responsible relation, a relation through which every being learns to respond and from which he learns to respond: it is through the grief one undergoes that life comes to matter; it is by accepting this grief that it counts. Taking the risk of vulnerability by facing up to grief so that vulnerable lives do not count for nothing, so that they "count as lives," assuming a becoming vulnerable together with and differently from animals, seems to me one way of responding to Haraway's proposition to create stories with companion species. This is what some farmers, whom Jocelyne Porcher and I have interviewed, are already doing: farmers for whom no decision is easy and who know grief firsthand. This is what speaks to us from the photos of some of their cows that hang on the walls of their homes, and it is similarly evident in the names they give to their animals—knowing that these very names anticipate a future sadness and the possibility of memory. And it is also what the farmers express when they maintain that they have not asked for forgiveness from their animals but instead say thank you.

Thinking like this does not make sense, at least no more than it does to honor deaths or to wonder what one honors, but instead requires that one search for meaning. And to learn to create it, even if it is not self-evident—especially if it is not self-evident.

L

FOR LABORATORY

What are rats interested in during experiments?

Vicki Hearne recounts how she once heard experienced experimenters advising young scientists never to work with cats. I would note, in passing, that it is also strongly discouraged, in laboratories, to work with parrots, not only because they never do anything that is asked of them but because they take advantage of their free time by destroying, with remarkable care, all of the equipment. According to the criteria of American experimenters, they are completely uncivil. An irrepressible curiosity, blatant boredom, or temperamental manifestation; all of these reasons can be invoked. In certain circumstances, Hearne continues (relaying the experienced experimenters), cats will complete a task quickly enough if you present one of them with a problem to solve or a task to perform to find food, and the graph that measures the cat's intelligence across comparative studies will show a steeply rising curve. But, and here she cites one of the experimenters, "the trouble is that as soon as they figure out that the researcher or technician *wants* them to push the lever, they stop doing it; some of them will starve to death rather than do it."[1] She adds, laconically, that this violently antibehaviorist theory has never, to her knowledge, been published. The official version has become, do not use cats because they screw up the data.

In contrast to what one might think, Hearne explains, cats do not refuse to please us. To the contrary. The expectations of humans are incredibly important, in their own eyes, and they take their tasks seriously. But it is precisely because it is serious to humans that cats refuse, simply because they are left no choice in whether to respond to these expectations.

All of this could have, for some of us, slightly vague hints of anthropomorphism to it. This scent is easily recognizable: in this story, the cat is credited with its own will, desires, and wish to collaborate, but not in any which way. One recognizes here the mark of a "non-scientist" (☞ **Fabricating Science**). In fact, before becoming a philosopher, Hearne was a dog and horse trainer; her desire to become a philosopher, she recounts, was motivated by the wish to find an appropriate way to translate the experiences shared by trainers and to find a language suitable to summarizing these experiences.

However, if she wants to claim, understandably, that saying the cats do not want to push the lever because they are compelled to do so is an undeniably antibehaviorist theory, then, practically speaking, she is not entirely right. If one follows the work carried out by the sociologist of science Michael Lynch, the behaviorists' quip about the failure of the laboratory—that "cats screw up the data"—has nothing unusual or strange about it. One hears much else besides in the laboratory. Lynch found that two perspectives coexist about the guinea pig in laboratories: the first is that it is an "analytical animal" and the other is that it is a "naturalistic animal."[2] This second perspective constitutes a body of tacit knowledge that is never mentioned within official reports but that is freely used during the course of actions, often in the form of comical stories. Humor, according to Lynch, places at a distance what cannot be inscribed within "doing science" [*faire-science*]. The two attitudes are at odds with one another, the second arising from a natural attitude that unfolds when beings endowed with intentions meet, the first responding to behaviorist demands to deny all possibility of contact between the experimenter and the object–subject. Consequently, the anthropomorphism that is constantly at work within the practice should disappear once one leaves the backstage and the scientist reports the results.

Saying "consequently," as I have just done, might be a bit hasty. It assumes that the anthropomorphism disappears through the simple effect of a scriptural translation, with the humor preparing the possibility of this withdrawal. But things are more complicated than this, because it begins well before the work of producing scientific articles. To begin with, the denial is not the simple product of an ascetic exercise of writing; furthermore, it does not only consist of denial; and finally, anthropomorphism is neither restricted nor absent from reports—*it is not perceptible*. In other words, it is made invisible.

This invisibility owes its efficacy to a series of operations and routines that accompanied the birth of the animal psychology laboratory.

These operations are mainly of two kinds. On one hand, practically speaking, the entire device is carried out in such a way that it blocks the possibility that the animal could show how he takes a position with respect to what is asked of him. In other words, the question "what could they possibly be interested in?" is never *seriously* asked. The researchers thus have an easy time not being anthropomorphic, as they neither let their animals yield to the temptation nor entice them to do so. On the other hand, if the anthropomorphism is not apparent, it is because the scientists invite us to focus our vigilance there where it is quite rightly easiest to control: in the writings and interpretations of the experimental reports.

"What could they possibly be interested in?" in fact constitutes a double question. On one hand, this question simply asks whether the experiment interests the animal. But because of the manner by which experiments are for the most part conceived, this first version of the question has no chance of being posed. The imperative of submission that guides the dispositives is at the very heart of this impossibility. There is absolutely no reason to question whether a hungry rat may or may not be interested in running a labyrinth and its branches, through which he must learn the route to discover food: he cannot do otherwise. The rat is not interested; he is motivated or incited. They are not the same thing.

That an animal actively resists or demonstrates his disinterest could certainly lead us to explore another possibility: maybe he is *not* interested? The solution is usually more simple: cats, parrots, and others will simply be excluded from learning. Most of the time, one says that they cannot be "conditioned." This is exactly what happened to parrots in the laboratories of behaviorists. Because they were unable to find a way to teach them to speak, the scientists who tried to do so ended up siding with B. F. Skinner, who asserted that language was instinctive and that one cannot condition instincts or reflexes, except for that of salivation—as Pavlov demonstrated with his dog and bell. It is not beside the point to explain how one has tried to teach birds, who were assumed capable of speaking, to speak: researchers placed parrots and myna birds inside test boxes and played a looped recording with words or phrases that, when heard, led to the automatic delivery of a food reward. Normally, according to conditioning theory, the subjects should learn to repeat the "conditioned stimulus." They didn't. The researchers thus concluded that Skinner was right. But the psychologist Orval Mowrer, while observing what happened after the experiment, remarked that one could have found a clue of *other* reasons: the assistants adopted two of the mynas as pets, and these ones spoke, quite fluently.[3]

Returning to the problem of "how can this possibly interest an animal?" another sense of this question is just as compromised by the way the device is conceived: one that explores how the subject of the experiment translates, in his own terms, his own way of being interested by the problem that is presented to him. Within this type of experiment, not only must the animal respond to the task that is addressed to him but above all he has to respond in the mode according to which the question is addressed. The assistants' mynas have never been the object of research or an article: they didn't follow the protocol, or if you prefer, they spoke for the "wrong reasons."[4] If an animal responds according to his own habits, in the register of what interests him, the researchers would consider this a kind of "ruse"—he admittedly did what was asked of him, but he did so for the "wrong reasons." The job of research consists, then, in flushing out these ruses and, to be sure, thwarting them. The case of speaking animals is exemplary in this regard: the use of recordings not only has the simple effect of a mechanization of work; they also "purify" the learning situation. If the animal learns with this type of device, he will be able to speak in any circumstance, and the act of speaking would not be dependent on a particular relation, with all of the influences and expectations of the researcher that "make speak" . . . In short, the competence would be abstract enough to allow any generalization.

The work to counter these ruses can take the most diverse forms, from the most mundane tasks to the cruelest mutilations. To limit ourselves to the least destructive versions, then, one learns that the scientists clean out, with great exaggeration, the labyrinths where the rats run. It had not escaped them that, after a few years of laborious work on theories of learning by conditioning, these sneaky critters were not memorizing which alleys ended in reward and which without; the rats were leaving a mark with their odor. These markings had nothing neutral about them, for they clearly indicated for each rat "this way is a dead end" (who knows, perhaps it was an odor of frustration?), "this way is a win."

With trickery of this kind, the rats were not providing evidence of learning based on memory but rather of something else that demonstrated the rats' talents—which interested neither humans nor the theorists. In other words, it doesn't matter how a rat might be interested in solving a problem posed to it, he must still solve it in the terms that interest the researchers. This in fact conveys the impossibility of the other version of the question "what could they possibly be interested in?" For one can see that if the animal responds by using his own way to arrive at articulating the problem, he no longer responds to the question "in general." Which means that his response has nothing generalizable about it. Even worse, if he responds for reasons associated with his relation to the researcher, or for his own reasons that have to do with the particular situation in which he is placed, then the "non-indifference" of this response compromises the process of generalization all the more. It is not that the given response is itself "indifferent"—in no way can it be—but the scientists feel entitled to think that the response resulting from the operation of submission is indistinguishable from all of the responses resulting from the same operations of submission.[5] It is there that the operation of submission takes on its essential condition: its invisibility. Every rat, in every labyrinth, runs because he is hungry. This is how the question is conveniently closed off. Provided, of course, that the apparatus has been cleaned. If not, you will have to consider other causes: the fact that the rat perhaps advances not because he is encouraged by the goal of food but rather little by little, each sequence following from the last, wherein he reads the messages and lets himself be guided by them: "not this way," "yes, there," "maybe a little farther," "I recognize this odor, I am on familiar ground." Other motives may also be at work; for instance, hunger might be forgotten so as to benefit other motives, as the rat reestablishes his own habits. Who knows what a rat's motives might be? Even worse, this was suggested to me one day by a young researcher: rats, she remarked, ran faster when there were spectators. This disastrous multiplication of possible motives is even further exacerbated if one refers oneself to Crespi, who claimed that rats, in certain experiments, modify their performances if they are disappointed by the reward or if they feel the elation of success when a reward exceeds their expectations (☞ **Justice**). Or worst of the worst, as demonstrated by Rosenthal's now famous experiment, rats learn their paths even faster if their experimenters think that they are more intelligent at the test, and the experimenters establish a much better relation with the rats if they're convinced the rats are intelligent.[6]

It's true that the risks of anthropomorphism are avoided when one avoids the difficult problem of elucidating, imagining, or considering the reasons why an animal might collaborate in research. To be sure. My first reflex, however, would be to deny this argument: anthropomorphism is always there, for what could be more anthropomorphic than an apparatus that requires an animal to deny his own habits to privilege those that the researchers think humans themselves do in the experience of learning? Except that researchers do not actually "think" that humans conduct the experiment in this way, and they don't even consider this; it isn't their problem. Their problem is that learning should be done for the "right reasons," which is to say for reasons that lend themselves to experimentation. As such, the particular form of the experiment's anthropomorphism is much more difficult to see. It comes in the character of "academicocentrism." This procedure of academicocentrism not only holds for the question of odors in the labyrinth, for it is even more discernible in the case of language learning. This learning rests on a narrow conception of language as a purely referential system that serves only to designate things, which is a very academic conception of language and through which learning arises only through memorization—which corresponds *grosso modo* to the way we study by heart. Neither humans nor animals learn to speak in this way. But humans can, through a long disciplinary process, in fact "learn" according to these procedures.

One could reproach me for having an inconsistent argument. I claim that a procedure that consists in washing a labyrinth or purifying a learning process of its relational elements is anthropocentric, and yet I concede to Crespi, with a sympathy that isn't hidden, that rats can be enthusiastic or disappointed by the fact that we respond—or don't—to their expectations. But I don't have any intention of being controversial. I am attempting instead to free the question of anthropomorphism from controversies by complicating it. In this context, complicating it requires that we reexamine the animals' habits and retranslate them into a perspective that could interest the animal and make it interesting for them, as well as interest ourselves (☞ **Umwelt**).

This question—"in what are they interested?"—leads to an exploration of more hypotheses and to speculating, imagining, and considering unexpected consequences, not in terms of obstacles but in terms of what obliges us. It is a risky question. It isn't mere speculation but research that is active, demanding, even cunning. It is a practical and pragmatic question. It isn't limited to understanding or revealing an interest but involves fabricating, twisting, or negotiating it with the animal. How does one make a parrot speak? What might he be interested in? One can no longer, of course, use behaviorist devices, with their recordings accompanied by food rewards. Other researchers, like Mowrer, understood the lesson: parrots need relationships . . . and a reward. But it was a lost cause. Mowrer's parrot only succeeded in learning to say "hello," and not even in the right way from the perspective of conversational standards. Receiving a peanut each time he said "hello," the parrot went on to imagine that "hello" meant "peanut." The relation was not enough, and neither was the peanut. Interest needs to be built. Irene Pepperberg initiated her research on this pretense. An interest has to be developed, twisted, even "tricked." She will do this with Alex, the gray parrot that she adopted from Gabon. First, a ruse that is well known among parrot handlers: these birds have a sharp sense of rivalry. So Pepperberg didn't attempt to teach Alex anything but instead asked him to attend some lessons that she would give one of her assistants.[7] At one point or another, the parrot would want to overstep the model. Alex spoke. And he would speak even more when he understood that, by speaking, he could obtain other things—things other than peanuts—as well as negotiate his relations with the team of researchers. And he did much more than this. I'm recounting this a little too simply, as if it were all self-evident. It was a long undertaking, both risky and demanding. Pepperberg assumed that what was in play had a twofold exceptional character: on one hand, because the language learned by the bird was that of another species, and on the other, because this learning process mainly took place during what one calls the "sensitive phase of learning," a period during which, under normal circumstances, a parrot will learn from his conspecifics. The term "exceptional" also implies, Pepperberg writes, that every possibility of resistance to this learning process must be taken into consideration with greater attention. This makes the exercise all the more demanding, as it requires cunning and tact, cunning and care. The attunement between tutor and student will be all the more subtle, all the more fitting: slowing down when it gets difficult, speeding up to avoid boredom, intensifying the interactions to ensure, she writes, "that the effects of learning are as close as possible to what would happen in the real world," namely, the ability to obtain things and influence others.

But *we are,* no matter what Pepperberg says, in a *real* world, the real world of a laboratory, as exceptional as it is, in which beings of different species work together—a real world in which every night a parrot says to his experimenter as she is preparing to go to back home, "Good-bye. I am going to eat now. See you tomorrow."[8]

M

FOR MAGPIES

How can we interest elephants in mirrors?

What could be the connection between, on one hand, Maxine, Patty, and Happy and, on the other, Harvey, Lily, Gerti, Goldie, and Schatzi? Very little. The former are Asian elephants around thirty years of age; the latter are magpies who are still young. The former live in the Bronx Zoo, the latter in a German laboratory. Their differences can look endless and predictable, but their commonality is surprising: they were asked to look at themselves in a mirror, and some of them seemed to be interested, with Happy on the elephants' side and Gerti, Goldie, and Schatzi on the magpies'. Harvey, for his part, attempted a few seductive maneuvers in front of what he believed to be a conspecific, became discouraged, revised his position in relation to the gender of the one who was facing him and mimicking all his gestures, and then simply attacked him. Lily was even more expeditious: she passed straight into aggression. After a few more attempts doomed to fail, the two magpies became disinterested.

On the first day, Gerti, Goldie, and Schatzi also obviously attempted to see if the "other" was actually a social being that would react appropriately. From the second day on, however, the three magpies interested themselves in another way. They looked behind the mirror (one can never be sure), they carefully examined the image in front of themselves, but they also found a decisive proof to solve the enigma: they would make movements that were a little unpredictable, like rocking back and forth, hopping up and down, or scratching themselves with one of their legs. One cannot be sure what these three magpies inferred from the situation, but clearly they understood that the other in front of them was not really an "other." There is still a leap to make to go from this to the claim that the magpies knew that it was themselves in the mirror. And one does not make a leap like this—not in the laboratory. One does not take a bird at its word, or by intuition, no matter how logical it may seem. This requires a test, a decisive one. The researchers—Helmut Prior, Ariane Schwarz, and Onur Güntürkün—therefore took it upon themselves to develop and propose one for the magpies.[1]

This test is well known today. It is based on the experiments that the psychologist George Gallup had proposed to chimpanzees at the end of the 1960s. The trial is simple—though a number of complications were added. After the chimpanzees became habituated to a mirror, the researchers would paint a red spot on part of their forehead while they were anaesthetized.[2] When they awoke, they didn't know about the presence of this spot. A mirror was then placed in front of them; if they searched for it on their own foreheads, then one could infer that they understood that the reflection corresponded to their own image. The test was simplified with the magpies, as the researchers decided to skip the anesthesia; in place of the paint, they substituted a little colored sticker (either yellow, red, or black) on their throats, just beneath their beaks, in a spot where the researchers could be sure that they could not see it, even when raising their heads. One of the three researchers covered the birds' eyes while another applied the sticker.

The procedure was a success: Harvey and Lilly, as expected in light of their past performances, did nothing with respect to the sticker; Goldie and Gerti, by contrast—and Schatzi too, though a little less so—actively tried to take it off, first with their beaks, then when that didn't work, with their legs. These magpies recognized themselves in the mirror. They were therefore self-conscious, or, according to Gallup's terms, they had a concept of self.

With the elephants, things are more complicated to piece together. It was all the more complicated because another researcher, a specialist in primatology, Daniel Povinelli, had already subjected two Asian elephants to the test a little while beforehand. They both failed. And yet they were capable of understanding a possible use of this device: in a preliminary test, Povinelli had hidden some food in such a way that the elephant couldn't see where, except through a mirror. The elephants had perfectly grasped its use, and thus the relation to the mirror was not thwarted by a visual deficiency. But obviously the task didn't seem to motivate them. The primatologist Frans de Waal, his student Joshua Plotnik, and Dania

Reiss, a dolphin specialist, attempted to stack the odds on their side.[3] What motivated them to resume an experiment that is clearly doomed to fail? Two of my students at the University of Brussels, Thibaut de Meyer and Charlotte Thibaut, looked into this question and carefully analyzed the articles and protocols from each of these experiments.[4] According to them, if Plotnik and his colleagues were ready to start over, it's because they relied on an observation made by Cynthia Moss, a specialist of African elephants. The elephants were capable of empathy; she'd observed many accounts of this. And if empathy correlates to the possibility of attributing mental states and desires to an other, she could therefore bear witness to the possibility of a theory of mind (☞ **Pretenders**). Povinelli, my students discovered, also mentions this observation. But he doesn't find it credible. It is only an *anecdote,* he says. There are, then, two types of approaches to knowledge that emerge from this contrast, and I'm not surprised that the contrast is modeled exactly after two different domains of research: Povinelli is a laboratory experimenter, whereas Plotnik and his colleagues are field researchers (☞ **Fabricating Science**; ☞ **Laboratory**; ☞ **Beasts**).

The three researchers next questioned why Povinelli's elephants failed. It's not impossible that a cause for this could be the mirror being too small or the fact that it was outside the elephants' cage, beyond the reach of their trunks. So they ensured that they offered a full-scale mirror for the elephants and that it was not outside of the cage but rather within it. Maxine, Patty, and Happy were faced with the mirror; during the preliminary test, they explored and even tried to climb up it, leading the guardians and researchers to be quite fearful that the wall against which it was leaning might collapse. And, like the three self-recognizing magpies, the elephants demonstrated behavior directed toward themselves: they watched themselves eating in the mirror, and they exhibited unusual repetitive movements with their trunks and bodies and rhythmic movements with their heads.

The day came to paint on the spot. Maxine looked at herself, touched the spot, and didn't stop touching it for the next several minutes. The other two didn't seem to want to pass this course. Two magpies and one elephant, therefore, clearly succeeded at the test, two magpies and two elephants failed it, and one magpie was reluctant. The experiment is an achievement [*une réussite*].[5]

One might be surprised by the way I just evoked this as an achievement. To qualify this, I'd compare Harvey, Lily, Maxine, and Patty's disinterest in the test to the clearly positive results of Goldie, Gerti, and Happy. For all of them, both "self-recognizing" and "non-self-recognizing" are important for this qualification. I speak of achievements because there were also failures. The possibility of failure, and what the scientists will do with this possibility, demonstrates the strength and interest of the experiment. If all of the magpies and all of the elephants had passed the test with success, the trial would not allow us to say what it can now claim for magpies and elephants: that they can be "self-recognizing." In other words, from the perspective of the researchers, the results of the experiment are much more convincing because some of the animals failed. As for me, I can only say it with as much conviction as I can, that the experiment is really interesting and that it makes the researchers, their magpies, and their elephants more intelligent.

To begin with, let's start with the obvious, with what we call without ambiguity an "achievement," that is to say, with what the researchers say about this achievement. I'll stick to the commentaries of magpie breeders, as they're quite surprising to hear in the general field of bird knowledge: "When magpies are judged by the same criteria as primates," write Prior, Schwarz, and Güntürkün, "they show self-recognition and are *on our side* of the 'cognitive Rubicon.'"[6] I believe that the metaphor of the cognitive Rubicon says what it means to say: that this story has an epic quality to it, with conquests and victories; there's an event, with a border that has been crossed and transgressed.[7] The die is cast: the magpies *(Pica pica)* will be the first birds to cross the border between those beings who can recognize themselves and those who cannot. But in this adventure, there is also a different story being spun: a story that, after some three hundred million years have elapsed since the moment of divergence between their taxonomic groups, reunites the corvids and primates—the magpies are now on *our side* of the cognitive Rubicon. After having thought for so long that humans were the sole custodians of the ontological treasure of self-consciousness, one came to accept that primates could claim access; next, through a contamination of talents that arises frequently enough in ethology, there came the dolphins, orcas, and then the three elephants, who beat out the magpies by two years in this story.[8] Until now, then, it has been thought that only mammals have had access to this competency. There would be, the authors also say in their introduction, "a cognitive Rubicon with apes and a few other species with complex social behaviour on one side and the rest of the animal kingdom on the other side."[9] This hierarchization, furthermore, received biological confirmation, because it came to correlate with the existence and development of the neocortex in mammals.

Let's return now to the fact that Harvey, Lily, Maxine, and Patty failed the test, which I'm trying to show is a sign for me of the achievement of the experiment. To begin with the authors, this failure did not imperil the strength of the results but rather confirmed them. Of the ninety-two chimpanzees who were tested by Povinelli in research before that with the elephants, only twenty-one showed clear evidence of self-exploring behavior in the mirror, nine showed weaker evidence, and, among the twenty-one animals who were "specular explorers," only half of them passed the spot test.[10]

However, if we dig a little deeper into these failures that I am calling achievements, I would want to emphasize a particular feature of this experiment that connects with what I call experimental achievements. This experiment, like others that resemble it, seems remarkable due to one of its features, which is immediately readable in one piece of evidence: it is an experiment of the culture of singularities. Harvey, Lily, Goldie, Gerti, Schatzi, Happy, Maxine, and Patty have nothing to do with the cohorts of anonymous beings that testify to the specificity of a species. This means that the failures of the non-self-recognizing animals not only signals the need to refrain from making generalizations: the experiment teaches us that magpies (some magpies, more specifically, magpies raised by hand) and some Asian elephants (roughly thirty years old and raised in a zoo) can, in some very specific and exceptional circumstances for magpies and elephants (☞ **Laboratory**), and developed with protocols (standardized and recounted with the utmost precision in the methodological section of the article), develop a new competency. But these non-self-recognizing magpies and elephants at the same time reveal the magnitude of this type of experiment. They are experiments of invention. The dispositive does not *determine* the behavior that is acquired; rather, it creates the occasion for it.

Now if *all* of the magpies and elephants had passed the test, this would indicate two possible things: either that behavior is biologically determined or it is the product of an artifact. The experiment, quite rightly, tells us nothing about the nature of the magpie or the elephant; it does not tell us that "magpies and elephants are conscious of themselves," it only tells us what the favorable circumstances are for this transformation. The competence arises neither unequivocally from the nature of these animals (the fact of being a magpie or elephant and not a pigeon is, of course, important, but if the competence were inscribed within their nature, they would all recognize themselves) nor from the effectiveness of the dispositive alone (this would have "forced" the magpies and elephants to recognize themselves); rather, it arises from the register of invention within particular ecological circumstances. Hence the importance of failure!

In other words, if all of them had succeeded, and the researchers had made a full assessment, one could always still suspect that the results were based only on an artifact. I would define the possibility of the artifact under the sign of success, as opposed to an achievement: yes, the hypothesis was validated and the experiment was a success, but it was only because the animal's adhesion to the hypothesis was the product of constraints imposed on the animal. To simply define this type of artifact, one could say that the animal responds to the researcher, but that he responds to a question that is utterly different from the one the researcher posed. So, to come back to our magpies, researchers will be wary of avoiding the possibility that animals are validating their hypotheses for reasons that only have to do with their obedience. Some researchers obtained rather similar behavior in pigeons when the pigeons were confronted with the spot test in the mirror. Yet, as Prior and his colleagues state, in analyzing the procedure, one realizes that the pigeons passed through an incredible number of conditioning tests, which resulted in producing the model behavior of recognition. The pigeons did what they were asked, but for entirely different reasons than the competence called upon; their response is a response to another question. It's worth noting that, in this group of subjects, and with this type of procedure, there is often an endless repetition of tests. It's therefore a risky precaution that scientists have to adopt with their magpies: they can only conduct a few tests, and the behavior must be, according to the researchers themselves, *spontaneous* and not the result of "blind" learning, which would not allow them to validate their hypothesis of a sophisticated cognitive competence.

The failure of Harvey, Lily, Maxine, and Patty thus translates the productive dimension of the test. The magpies and elephants who were enlisted in the experiment were able to resist the proposition that was made to them. Allowing these subjects to be "recalcitrant" opens the dispositive to the element of surprise because it submits it to risk." There were hardly any risks with the pigeons: they are among the best peddlers of the efficiency of conditioning. All of them had the expected reaction in front of the mirror, once it had been taught to them. But it was a high price, because the researcher demands the autonomy of produced facts. The dispositive had totally determined them.

The failure of Harvey, Lily, Maxine, and Patty therefore signals an achievement par excellence. The autonomy of produced facts—what these scientists call "spontaneity"—conveys the fact that the dispositive is a necessary but not sufficient condition for their production. Of course, without a mirror, without work, without taming, without a spot, without tests, and without any observations, no magpies or elephants would be self-recognizing; however, if the magpies or elephants are compelled by the dispositive, then this evidence could not permit a distinction from those animals who are conditioned. The difference may be expressed through a term that is likely poorly defined but that nevertheless leaves open, through its various homonyms, a large repertoire of habits and speculation—that of being *interested*. We don't know what might have interested these self-recognizing magpies and elephants in the test, and there could be many hypotheses. But this question is just as interesting if it's inverted: why aren't the non-self-recognizing animals interested? The act of posing it expresses, on the part of the researchers, forms of consideration that are both epistemological and ethical, in the rich etymological sense of ethos, namely, that of "habits" and "good habits." For the elephants, therefore, they believe that the reasons for their failures might have to do with their habits. A spot may not move them at all because their habits in matters of cleanliness are not the same as those of birds or chimpanzees; their washing habits do not consist of removing dirt but rather of throwing mud and dust over themselves, without much attention to the details. So, a little spot in all of this . . . Furthermore, the comparative analysis of dispositives carried out by de Meyer and Thibaut raised an important difference between what was proposed by Povinelli, on one hand, and by the teams of de Waal and Prior, on the other. With the latter, the animals could touch the mirror. According to my two students, these animals could build an emotional rapport with the mirror. I'm not sure that this is the term I would use, but it does open itself to another, which falls within the area of what can *affect*: they could allow themselves to be *affected*. Because the scientists were paying attention to their habits, these animals could "play" with the object, which is to say they could invent—in an imaginary, exploratory, affective, sensitive, and concrete way—very different habits. And it is by multiplying and inventing these habits that what is proper to them may well have crossed paths with our own. For mirrors, don't forget, bring us back to ourselves. One cannot say if, in recognizing their reflections, these magpies and elephants encountered themselves; but they certainly have tangled us up.[12]

N

FOR NECESSITY

Can one lead a rat to infanticide?

"Among polygynous species, a growing number of observations are showing that when a male takes possession of a harem from an evicted predecessor, he sometimes kills the infants, which accelerates the estrus of females and allows him to copulate. These infants would therefore be the bearers of his genes."[1]

This kind of statement is widely reported today, not only in the scientific literature but also in popular books and various documentaries of "the weird world of animals" kind. It appeared at the end of the 1970s when some researchers found themselves confronted by some problematic observations: among some animals, the adults kill the infants of their conspecifics. The explanation of this problematic behavior as "adaptive," however, was quickly imposed and remains still dominant today. Upon reflection, the impression that the quoted description gives is far from obvious. It passes carelessly from observations to a biological explanation, attributing as a fact what is in principle no more than a hypothesis of a cause. This shortcut in the chain of translations is not just a simple result of popularization but signals the fact that infanticide is inscribed within the domain of biological necessities.[2]

The question of infanticide arose as a result of observations made of some monkeys, the Indian langurs, who were at the time relatively obscure. In contrast to gorillas—who were much more popular, albeit much more rare—almost no one knew about the langurs. The problems posed by these obscure monkeys nevertheless captivated the public. This captivation would be less surprising if it were put back within the context of the time. The incidents observed among the langurs, and the theoretical proposition that will make sense of them, coincides with the moment when domestic violence and, more specifically, child abuse suddenly arose as real social problems.[3]

It can be said with respect to a number of animal behaviors that fieldwork often opened the path to the laboratory. What researchers in the former area report—and what is often held with a bit of disdain by the latter (that they are nothing but anecdotes)—will, at one point or another, find themselves subject to the scientific test par excellence, the experimental test. The laboratory constitutes the ascension of observations and provides a miraculous transformation of anecdotes into scientific facts (☞ **Pretenders**; ☞ **Beasts**). Infanticide will go through a swift ascension. Beginning in the mid-1980s, articles emerging from experimenters will multiply. This is likely related to the fact that this behavior was similar to human social problems of the same period—and the fact that rats were introduced within this story comforts me in this impression. Rats, who until now had tested every drug possible, who were plied with alcohol and cocaine, who ran labyrinths for the behaviorists, inhaled thousands of cigarettes, knew experimental depression and neurosis, learned how to measure time . . . these loyal servants of science had thus become infanticidal![4]

They must be given credit: as will be seen, rats are not particularly adept at this kind of behavior—but then again, they're not particularly happy about smoking cigarettes, testing drugs, or running labyrinths while hungry either. If they were called upon for this kind of research, as with the others, it's because they were the most convenient laboratory animal—relatively economical, easily replaceable, and, undoubtedly, the most manipulable among experimental animals. Rightly or wrongly— they were not asked for their opinion—rats would become infanticidal.

The scientific literature informs us that this behavior can be exhibited by the mother, an unfamiliar male, or an unfamiliar female—one should probably add to this list of culprits the researcher and laboratory technicians who are in charge of euthanasia when the young rats become too numerous, but this would spoil the publications. In terms of the mothers, it has been found that they might kill their infants when they are born with malformations, when they are stressed and see their environments as hostile, or, further still, when they are starving, which leads them to eat their infants. In terms of the males, one finds the hypothesis that by killing the young, the male promotes the return of the female's estrus and thus allows him to breed more quickly. And yet, the researchers explain, one will find that infanticide is inhibited if a male has been in the presence of the female during the gestation period or if he is frequently put in the presence of infant rats, which elicits parental behavior. As for the females unfamiliar to the mother, the last category of the potential culprits, they practice infanticide to feed themselves or to take possession of the female mother's nest. One observes, by contrast, that when females are raised together, not only is infanticide rare but they help one another in caring for the infants.

If one sticks to the conditions for the emergence of this behavior, one realizes that the conditions that are said to be able to "reveal" infanticide appear as, above all, conditions that are actively created by the researchers. Who had the idea to starve the rats? Who had the initiative to put unfamiliar males into a cage and in direct contact with mothers who had just given birth? Who organized the distribution of cages to put females unfamiliar to one another side by side—and most likely with only enough material to make a single nest? How did the environment become stressful and hostile? One cannot ignore that these are extreme conditions of captivity, that is, conditions of experimental captivity manipulated so as to induce stress, hunger, hostility, fear, and so on. In short, these are pathological conditions, carried to the extreme, that clearly have the goal to force the behavior; the researchers will repeat and vary the test until the desired behavior appears. We are dealing with a tautological operation: infanticide is the behavior that emerges when all of the conditions that induce infanticide are met so as to make it emerge! The next leap consists in considering that these conditions have an explanatory value. This leap is quite noticeable when reading the articles.[5] When the researchers report the circumstances in which infanticide did not occur, one can read, as they put it, that *there are conditions that prevent infanticide*—not conditions in which infanticide does not arise but rather conditions that neutralize it. This means that infanticidal behavior, just as much as "noninfanticidal" behavior, is actively induced, because the behaviors do not emerge in the absence of the conditions that give rise to them. One can conclude only one thing from this: the researchers ended up thinking that infanticide is the expected, and thus normal, behavior, whereas noninfanticide is the behavior that needs these conditions to come about. Strange inversion. The exception becomes the norm within experimental conditions, and what should normally happen instead becomes the exception. Temple Grandin, a specialist in livestock animals, would probably oppose this with the succinct judgment she uses when breeders are not moved by the fact that their roosters violate and kill the hens or when llamas bite off the testicles of their companions: "That's definitely not normal."[6] If it were normal, she says, there would no longer be any roosters or llamas in the wild. This reasoning can be extended to rats who feed themselves on their descendants.

This inversion of the normal and the pathological describes what happened in the laboratory: the researchers act as if they were only revealing what preexisted their research, for they do not take into account that infanticide actively receives its conditions of existence from the dispositive, that it results from the work of fabricating necessary conditions, a work that has effaced some results. This in turn authorizes the experimenters to assert the possibility of generalizing beyond the laboratory: here are the conditions, *in general,* that cause infanticide, and here are the conditions, *in general,* that inhibit it.[7] Infanticide has become a spontaneous, "natural" behavior, especially if one conceals that nature is laboriously fabricated within these dispositives. The evidence being, for infanticide to be absent, it requires the abstention of the procedures that created it.

Obviously, this does not mean that there isn't any infanticide in the wild. It's because it was observed there that this research was even carried out. Returning to the field, at the end of the 1970s, the first observations shocked and intrigued researchers. The theory that I have just evoked emerged incredibly quickly: the male commits infanticide; by killing the infants of another male, he appropriates the harem; he promotes the return of the female's estrus; and he can thus impregnate and propagate his genes.

This explanation is based on the sociobiological theory of intrasexual competition that reports the strategies adopted by one sex or the other in relation to their rivals during the course of reproduction. It has been formulated with respect to lions, gulls, great apes, langurs, and many more species. This theory is marked by a form of maniacal obsession whose major symptom is a shocking tendency to stereotype. All forms of behavior are passed through the same sieve: that animals have but one preoccupation in mind, to assure the dissemination of their genes. Their existences are bound by the strict confines of necessity; not only is nothing without selective cause but a single cause prevails, and it is one that comes from the general scheme of *adaptation* strategies. There is no question of indulging certain motivational extravagances like the act of singing, grooming, playing, copulating, or watching a sunrise for the simple pleasure of doing so, or because they are the social habits of the group, or because status, bravery, or relationships are important. To cite but one example, some primatologists observed that, within a troop, some female chimpanzees mated with *all* hypersexualized males present. They concluded that . . . it is a strategy to avoid infanticide, because every one of the males could have had the chance of being the father. There you have it, a perfectly virtuous translation: sexual depravity in fact proves to be evidence of wise maternal foresight . . . This kind of hypothesis indicates the complicity of reflexes arising out of the natural sciences with the chauvinist and Victorian prejudices concerning the sexuality of females. Searching for the usefulness of behavior—a kind of bourgeois moral of evolution that does not waste its time with useless antics and that leads researchers to search for the adaptive value in every behavior—avoids all the hypotheses that do not support the idea of a long-term selective interest, such as the hypothesis of pleasure, the strength of the drives, or a simply extroverted sexuality. In terms of the males, the latter hypothesis would be accepted—except that we would be told that they are ensuring their lineage—but in terms of a female, do not even think about it.

Returning to the langurs, the first observation that opened the way to re-search on the subject of infanticide was reported by a Japanese researcher, Yukimaru Sugiyama, while working in India. Infanticide occurred while important social changes were transpiring in a troop. To be more specific, these social changes were due to the initiatives of the scientists. They carried out an "experimental manipulation" of the troop. Sugiyama trans-ferred the only male of the group—a male whom he said was the dominant, sovereign male and who protected and managed the harem—into another group, which for its part was bisexual. Also worth specifying is that this type of practice was common among some primatologists, more specifi-cally among those who seemed to be especially fascinated by hierarchy (☞ **Hierarchies**). Following this experimental manipulation, according to the very terms of Sugiyama, another male entered the troop from out of which the other male had just been removed, took possession of the harem, and killed four infants.

A little later, another researcher, the sociobiologist Sarah Blaffer Hrdy, observed infanticides perpetrated by some male langurs, also in Jodhpur.[8] She reinforced the thesis according to which the male manipulates the estrus cycles of females through enforced infanticide so as to ensure the perpetuation of his genes. Also notable is that at the same time another researcher, Phyllis Jay, was similarly working in the field with langurs, in another region of India. She didn't observe anything similar to this. But she would go on to comment on the other research, which I'll come back to.

I would like to dwell for just a moment on the manner in which Sugiyama's observations were formulated. The semantics used are not innocent: not only does it describe certain things, certain theoretical biases, but it also encourages the choice of certain meanings. Evoking what has occurred by talking about the male who "takes possession of the harem" and who replaces a "dominant sovereign" in order "to protect and manage the harem" himself—I am simply lining up Sugiyama's semantic choices, who is himself adopting the terminology in practice—already engages in a certain type of storytelling.

The issue is therefore not about critiquing the words used but about working within a pragmatic perspective. What type of narrative does this kind of trope engage? Or, more concretely, can one reconstruct the story by using other tropes? Would different words not make this story less obvious? For instance, the term *harem* usually refers to a group composed of one male mating with several females. This semantic choice implies a particular scenario: that of a dominant male exercising control over his females. However, who says that the male chooses the females? That he appropriates them, that he takes possession of them? Nothing does; it is only the term *harem* that encourages this meaning.[9]

Another way of describing this type of organization has been proposed, however, notably by some feminist researchers working in the framework of the Darwinian theory of sexual selection—according to which it is the females, in most scenarios, who choose the males. To describe this type of polyganous organization, the researchers proposed the following scenario: if one male alone is sufficient to ensure reproduction, and males pay little attention to infants anyway, then why would the females want to deal with several of them? If one is sufficient, and is capable of holding the other males at a distance, the females therefore have every interest in choosing a unique male rather than burdening themselves with other individuals. So there you have a completely different story than that of the harem, a story that holds up just as well and that proves consistent with the Darwinian perspective.

But this story doesn't simply reverse the perspective of narration, it obliges us to change the narrative structure itself. The story that describes the effects of relocating the male no longer has any evidence to support its circulation; and nor does it simply consist of an unknown male who imposes himself, takes possession, and manipulates the females' estrus through imposed infanticide.

Another story can therefore begin to be imagined—a story that has a double merit: it complicates the problem by pulling it out of the un-equivocal and monocausal register of necessity, and, in doing so, it no longer puts the entire burden of explanation on infanticide. Obedience to an imperious biological necessity, then, is no longer the motive or cause but could turn out to be none other than a side consequence of other things, things that require that one pays attention, which an "all-terrain" hypothesis saves us from.

The credit for opening this alternative scenario for the langurs is owed to Phyllis Jay. As I mentioned earlier, Jay was studying langurs in a different region of India. She hadn't observed any infanticide, but her knowledge of the animals involved would lead her to take part in the theoretical de-bate. She analyzed the field data from where these events were observed. She took into account the experimental manipulations and, for the non-manipulated groups, the context within which the observations had been collected. A careful analysis of the theories, the semantic choices of her colleagues, and what happened to the langurs led her to conclude that it is much more relevant to consider infanticide not as a strategy but as a consequence. On one hand, she says, infanticide shouldn't be understood in the context of a takeover because this is a term that too strongly dictates the narrative. It is there, as Haraway reminds us (and too whom I owe a large part of what has guided my present reading), that we see that words and the manners of speaking are important and that there is nothing innocent about them. Narrative structures keep the attention on some things, while inhibiting it from others. While one focuses on this story of the harem and conquest, one doesn't pay attention to what has occurred as a consequence of the experimental manipulations, like *the fact that the only male of the troop was a victim of kidnapping*. Perhaps he was sovereign, but what does being sovereign mean? Evoking deference, emotional ties, creating an atmosphere of trust? If langurs have different options, which they clearly do, because they can live in groups that are bisexual or polygynous, and if the theory of choice for females is right—that they create very particular attachments to this male as opposed to that one—then one can imagine the trauma to the group. "Our male was taken away by humans who constantly observe us." Anything and everything can then happen. Within this framework, the causes of infanticide become much more contextualized. They oblige us to consider the fact that a society is built day to day, that it composes itself, and that it can at any moment unravel if some irresponsible humans interfere. Jay's analysis of nonmanipulated groups converges with this one. Observations of these groups allow one to infer that infanticides occurred when social changes were taking place too quickly within contexts with population densities that were too high, that is to say, within highly stressful conditions that are themselves sufficiently pathogenic. A good number of observed infanticides, she goes on to note, are in fact accompanied by the killing of females, so the uncontrolled aggression of the male is not directed against the infants alone. Infanticide is not an adaptation but rather the sign of a *disadaptation* to contexts that are too new and too brutal.

Jay's explanation did not take hold, however, and the sociobiological theory, in this regard, remained the dominant explanation. Within scientific circles, sociobiological theory seems to have, to all appearances, succeeded in the conversions it sought, as evidenced by its impression of being self-evident, which I've highlighted from these popular writings. This was not entirely the case, however; the controversy never succeeded in being resolved. Some resistance followed that of Jay. After a bit of a lull—which is usually the sign of the end of a controversy—a primatologist, Robert Sussman, reopened the debate. He dissected the specific context of each reported case of infanticide within primatology research. In his analysis, infanticide attacks were far less common than previously estimated, with just forty-eight reported. And from among these forty-eight, nearly half of them occurred at the Jodhpur site. Moreover, of these forty-eight attacks, only eight aligned with the predictions of the adaptation hypothesis. Is it thus realistic, he asks, to consider events that are so rare, and that for the most part seem confined to a particular spot, to be exemplary of an adaptive strategy? Moreover, it is notable that at Hrdy's field site in Jodhpur, some German researchers were themselves observing the langurs in the early 1980s, and they never witnessed the case of a violent infant death.

In the case of lions, another scientist, Anne Dagg, took up the same method toward the end of the 1990s. All of the research on lions had until that time advocated for the theory of sexual competition. Dagg stated that, in reality, no case of infanticide could correspond to a "typical situation" capable of supporting this adaptationist theory. Her research triggered quite a bit of anger and hostile reactions from among her colleagues. Jay herself would return to the debate at this time with an article in which she shows that the langur infants are in fact quite interested by the conflicts between adults. Often the accidents would not be due to the fact that they were the targets of overly excited adults, as previously thought, but because they were "in the midst of it all."

As the sociologist Amanda Rees has remarked in her retracing of this story, it is rare for a controversy not to find its resolution, at one point or another, in the field of ethology. In this respect, then, the case of infanticide appears to be rather distinct: it never stops being the object of questioning, for every time the issue is believed to be settled and resolved, a scientist will refuse the "conversion" to the sociobiological theory and will relaunch an investigation. This impossibility of closure is all the more surprising given that the observed cases ultimately prove to be so rare—and even tend to become more rare as the analyses begin to reinitiate the debates. It's true that, as I've highlighted here, the problem stems from political issues; it's directly tied to serious human problems for which the ways of explaining and responding to them are themselves controversial. It could have been remarked from the outset, as Rees has emphasized, that interpretations made from the field have been the object of suspicion for political interference in science. The very fact of considering infanticide as an adaptive strategy, as can be read in the way that sociobiologists have described these situations, arises from a chauvinist ideology. However, if we rely on this political argument, shouldn't it also be asked if, as the sociobiologists have indeed claimed, the willingness to consider infanticide as an accident isn't itself a moral judgment on nature—"in principle this shouldn't happen"?

I'm not at all sure whether these types of arguments are helpful, at least not in these deconstructive or critical ways; they participate in the controversy, to be sure, but deconstruction misses the most important issues. These are two different ways of doing science, two ways that are at stake and in tension with one another in the area of animal studies. On one hand, we have a method inherited from biology and zoology that looks for similarities and invariants, within species and more generally between species, by requiring animals to obey laws that are susceptible to generalizations and to relatively univocal causes that can be inscribed within an interpretive routine. At the same time, this practice resituates, so as to perpetuate, a customary habit of laboratory practices: that of constructing, in the field, the repeatability of events (by considering all events, in the context of the field, as identical), just as in the laboratory one submits to the obligation to repeat experiments. This requirement is based on the conviction that all contexts are ultimately equivalent. It is a method that requires the submission of nature—just as the laboratory requires the submission of its subjects—in doing science (☞ **Laboratory**). On the other hand, we have another practice that is in competition with the former; this one is inherited, meanwhile, from anthropology's ways of thinking and doing and seeks to explore, by focusing on their flexibility, the singular and concrete situations encountered by animals that describe every event as a particular problem that animals are experiencing and at-tempting to handle (☞ **Reaction**). It is just as much about politics, but the politics of science and of political relations to nonhumans.

In addition, if all milieus are a priori equal for the first group—the sociobiologists—and adaptive strategies and programmed patterns over-determine behavior, then the second group, in contrast—which further indicates the fact that they are the heirs of anthropology's methods—have taken into consideration that the very processes of industrialization and globalization that allow them to travel and carry out their field research in faraway places are exactly the ones that confront their animals. These processes affect and considerably alter the lives of animals, whether through destruction of habitat, tourism, or urbanization. It is not a matter of denying that these animals, like all living beings, cope with biological necessities but rather of actively considering the very conditions of their concrete existences, in the sense of noncausal conditions, in the sense of what makes their lives what they are. Lives that are, now more than ever, and for each of these animals, *with us,* lives with whom we have a role in their vulnerability. And it is also in this sense that the problem of infanticide is a political problem.

O

FOR OEUVRES

Do birds make art?

Can animals create works of art?[1] The question is not far off from the one that asks whether animals can be artists. To test this, at least speculatively, raises again the question of intention, which in principle should preside over any work. Must there be an "intention" to make a work, and if there must, is it the intention of the artist that determines, or not, whether she is the author of the work? Introducing animals into the posing of this problem has the merit of making us hesitate and slow down. Bruno Latour has made us sensitive to these hesitations by proposing a reconsideration of the distribution of action in terms of "making one make" [*faire-faire*].[2]

It is worth considering the splendid arches of the pink-naped bowerbirds, which are all the more interesting because these birds have reappropriated, for the sake of their own works, some of our artifacts and put them to use in their compositions. If one pays attention to the work accomplished—one merely has to enter the name of the bowerbird in any search engine—one can see, thanks to the camerawork of biologists, that there is nothing accidental about the composition; it is all organized to create an illusion of perspective. According to the biologists, it is all staged to make the bowerbird dancing in his arch appear larger than he actually is. We are therefore dealing with a scene, a staging [*une mise en scène*], and a truly multimodal artistic composition: a sophisticated architecture, an aesthetic balance, a creation of illusions designed to produce effects, and a choreography that concludes the work—in short, what the philosopher Étienne Souriau would likely have recognized as a *poetry of movement*. This skillfully orchestrated illusion of perspective refers us to how he proposed to make sense of simulacra. They are, he writes, "sites of speculation on meaning" that clearly testify to the capacity in nature to *create being out of nothing in the desire of the other.*

Creating being out of nothing in the desire of the other: is this a work of art in the sense that we understand it, and for which the bird would be the true artist and author (☞ **Versions**)? I am temporarily leaving to the side the sterile and boring debates that attempt to reduce the animal to instinct (☞ **Fabricating Science**) and that, to provide an account of the work accomplished, provide us with explanations of the causally deterministic and biological kind. It is also worth noting, just in passing, that in terms of these kinds of explanations, sociobiologists have similarly tried to apply them to humans: every action and every work might be explained by a program to which one is genetically bound and whose goal is to better perpetuate one's genes (☞ **Necessity**). I leave it up to the reader to describe this in less carefully chosen terms. The fact that these explanations are in such bad taste and so impoverishing ought to prevent us from using them with nonhumans who have already been so mistreated by theory![3]

On the other hand, I could take up the way that the question is posed by Alfred Gell, the anthropologist of art, when he asked it not about animals but about artistic productions in cultures that do not consider their productions artistic.[4] Gell's problem is the following, albeit summarized a bit quickly: if one considers art to be what is received and acknowledged as such by the institutionalized world of art, then how should one consider productions from other societies that *we* consider as artistic, whereas these societies do not themselves accord the same value to these objects? To not do so, as has been the practice for so long, would return the others to a status of primitives expressing their primary needs in spontaneous and childlike ways. To do it anyway, as Gell explains, obliges the anthropologist who is studying the creation of objects in other cultures to impose on these cultures a completely enthnocentric frame of reference. Indeed, if one considers that some of the objects do not have any aesthetic value for either the ones who produce them or for whom they are made, then the solution that consists in placing each production in the cultural framework of the one designating the rules and criteria of aesthetic taste does not solve the problem. Put more simply, a shield, for example, is not art for "them" but for "us."

How to escape this impasse? Gell proposes that the problem be redefined. Anthropology is the study of social relations; one must also therefore consider studying the production of objects within these relations. To avoid falling back into the impasses that I have just recounted, however, the objects themselves ought to be considered as social agents endowed with the characteristics that we give to them. Gell, therefore, attempts to take the question of intentionality out of the narrow framework in which our concept has confined it and instead open up the notion of the agent—as a "being endowed with intentionality"—to others besides human beings.

A decorated shield, to take up again the problem of objects that carry for us an aesthetic value, doesn't have this value in the context of a battle in which it is used. It elicits fear, or fascinates, or captivates the enemy. It signifies nothing and symbolizes nothing; it acts and reacts, it affects and transforms. It is, therefore, an agent, a mediator of other agencies. The concept of agency (which the French translator of Gell's *Art and Agency* translates as "intentionality" [*intentionnalité*]) is therefore no longer posed as a way of classifying beings (those who would be ontological agents, endowed with intentionality, and those who would be ontological patients, devoid of intentionality). Agency (or intentionality) is relational, variable, and always inscribed within a context. The work not only fascinates, captivates, enchants, and traps the recipient; rather, it is the agency contained within the very material of the work to be made that controls the artist, who thus takes the position of patient. If I understand Gell in Latour's terms, the work makes happen [*fait-faire*]; the shield makes the artist make (the artist *is made-to-make* by the shield), it makes the one using it (e.g., it can make one more daring in battle), and it makes the enemy warrior (e.g., be fascinated, scared, captivated by it).[5] In our relations to artworks, Gell says, we are quite similar to how the anthropologist Edward Burnett Tylor described the indigenous peoples of the Antilles: they claimed that it was the trees that called to the sorcerers and gave to them an order to sculpt their trunks in the form of an idol.

By distributing intentionality in this manner, Gell somewhat agrees with what Souriau proposed, albeit with much more speculative prudence. According to the latter, the work imposes itself on the artist, or if I were to use Gell's terminology, "it is the work that is the agent," it is the work's intentions that "insist," and it is the artist who is the patient. Nevertheless, if I now want to ask about the possibility of art among animals, and do so seriously, I must abandon Gell and align myself with Souriau. For even if Gell clearly redistributes intentionality and agency, he reduces the redistribution, despite a few worthy attempts, to a relation between the work and its recipient. He writes, "Anthropologists have long recognized that social relationships, to endure over time, have to be founded on 'unfinished business.' The essence of exchange, as a binding social force, is the delay, or lag, between transactions which, if the exchange relation is to endure, should never result in perfect reciprocation, but always in some renewed, residual, imbalance."[6] He continues, "So it is with [decorative] patterns; they slow perception down, or even halt it, so that the decorated objects is never fully possessed at all, but is always in the process of becoming possessed. This, I argue, sets up a biographical relation—an unfinished exchange—between the decorated index [which means the 'work' object as carrier of intentions] and the recipient." In short, the speculative leap that distributes the intentions between the work and artist is not carried through to its end, for Gell clearly hesitates to make Antilleans of us, a sorcerer of the artist, and a summoning agent of the work.

This question is posed entirely differently by Souriau when he evokes, in his 1956 paper "From Modes of Existence to the Work to Be Made," and in terms that appear to be similar, the existential incompleteness of everything.[7] But the incompleteness of the work, for Souriau, is not found between the work and its recipient but rather between the work to be made [l'œuvre á faire] and the one who will devote herself to the work, the one who "must respond to it," the one held responsible. Works to be made are real beings, but whose existences demand promotion on other planes. They are deficient in existence, if only because they only benefit from a physical existence. A work, in other words, calls for its fulfilment on another mode of existence.

Can we return to the problem of animal artists with what has been proposed here? Souriau anticipated this question with his book *The Artistic Sense of Animals*.[8] From the very first pages, he evokes the sense that his response will take: "Is it really blasphemous to think that art has cosmic foundations and that one can find in nature the same great *instaurating* [instaurateurs] powers?" The term "instaurating" is not chosen by accident. Souriau did not use "creator" or "constructor" (even if he at times considers these terms as equivalent, we are still well before the arrival of constructivism, so "construct" is not yet a loaded term). Instaurating means something else.[9]

The work, as we've just seen, *calls for its accomplishment on another mode of existence.* This accomplishment requires an instaurating act. In this sense, if one can say that the creator *carries out* [*opère*] the creation, the being of the work nevertheless exists before the artist has made it. However, this being could not have made itself by itself. "To insaturate is to follow a path. We determine the being to come by following its path," he writes. "The being in bloom," he continues, "reclaims its proper existence. In all of this, the agent has to bend to the work's own will, to divine its will, to abdicate himself for the sake of this autonomous being that he is seeking to promote according to its own right to existence."

To say that the work of art is insaturated, then, is neither to attribute causality somewhere else nor to deny it. It is to insist on the fact that the artist is not the cause of the work and that the work alone is not its own cause; the artist carries responsibility, the responsibility of one who hosts, who collects, who prepares, who explores the form of the work. In other words, the artist is responsible in the sense that he must learn to respond to the work, and to respond to his accomplishment or his failure to accomplish such work.

If we return, then, to our question, can we imagine speaking about natural beings as masters of a work? To be sure, when Souriau engages with this question in his book on the artistic sense of animals, he seems to hide at times behind a form of vitalism that is particularly noticeable in the commentaries that accompany the images: "Life is the artist, the peacock is the work." For that matter, however, in returning to the birds, one discovers this surprising proposition beside a photo showing a zebra finch in the process of making its nest: "The call of the work." Here, quite clearly, it is no longer a matter of an abstract nature but rather of an instaurating being, responding (as the one responsible) to the challenging demand of accomplishing a work. Beneath this title, Souriau explains that "often the nest is made by two of them, and its preparation is essential to sexual courtship. But occasionally a celibate male will begin this work alone." A female could join him and help, he says, and it is in this sense that the nest is a work of love or rather, as he corrects himself, "a creator of love: the work mediates."

Invoking love the way he does makes me want to prolong it. The work really has the power to captivate those who carry out its accomplishment. It is thus a completely different theory of instinct that we are invited to consider. It is a theory of instinct that, far from mechanizing the animal and returning it to biological determinism, instead offers, in a speculative mode, much more fruitful analogies.

Let us return for a moment to the nests of the bowerbirds that I raised earlier and take up again the question where it was left off, somewhat entangled between instinct and intentionality. I am not responding to the question of knowing whether these birds are artists, for it is no longer here that the problem interests me. If I were to go back over one of Gell's examples, namely, that of the shield, then by following the analogy, one could maintain that these nests are objects that captivate, transform, and produce beings that fall in love or that they enamor, fascinate, and have an effect on them. But if I follow the path opened by Souriau, and interest myself not in the relation with the recipient but instead with what deploys the instaurating act of the nest, then I could also suggest that the pink-naped bowerbirds are well and truly captivated by the work to be made and that it is really this that dictates the work's need to exist. "This must be."

Of course, our preferences tend to instead favor the idea that a work can only be made by a special few who are less dispersed, for this is how we consider art, with a kind of exceptional status. It is without doubt this lack of exceptionality that justifies the cumbersome recourse to the argument: if anyone can do it, it must be instinct. It's true that for these birds, the making of this work of art is tied to vital questions, because the making-of-the-work [le faire-œuvre] is for each bird the condition of its preservation. Without the work, there will be no descendants who themselves will make future works. But do not confuse a condition of preservation with a condition of existence, and do not confuse what the work makes possible with its motive. Or, at any rate, abandon the concept of instinct, but guard preciously what it makes us feel, what feels like a force in the face of which being must bend—like we sometimes do in the face of love. No matter what utilitarian aim we might impart to these works, we know that birds do not have this utilitarian aim in mind (the motives are always identifiable a posteriori, a convenient rationalization that is pertinent from a biological point of view but might not be what the birds say is important). What instinct both affirms and masks is the call of the thing to be made. That some things are beyond us. The captivation known to some artists. That this must be made. Period.

P

FOR PRETENDERS

Can deception be proof
of good manners?

A monkey that was fond of climbing a tall pole outside had taken up the habit of remaining up at its summit. Yet each time that he was brought his plate of food, some crows in the vicinity would immediately come to steal it away. This scene repeated itself every day; and every day the poor monkey had no other choice but to give in to the incessant coming and going from summit to ground each time that a crow shamelessly approached for his pittance. As soon as the monkey approached, the impertinent birds flew off, only to land a few meters away. The monkey would climb back up, and the crows would return. One day, the monkey showed signs of a debilitating sickness. He was in such a miserable state of dejection that he could barely keep hanging on to the pole. The crows, as was their habit, came with utter impunity to take away their share of the meal as the ailing monkey, in a good deal of trouble, would come down the pole. He would finally let himself fall to the ground and would remain there, sprawled out, without moving, in obvious agony. Reassured, the crows became bolder and would quietly return to accomplish their daily theft. Then suddenly, the monkey seemed to miraculously recover all of his powers and in an instant jumped on one of the crows, trapped it, clasped it in his legs, plucked it vigorously, and threw his victim, who was as stunned as he was featherless, into the air. The result was the whole point of his act: no crows would ever venture around his plate again.

This story was written by an author who is not at all contemporary: Edward Pett Thompson was in fact a naturalist from the early nineteenth century.[1] And he was a creationist. In reading it, however, one can't help feeling a sense of familiarity. It resembles those stories that are produced today by scientists working with some of the animals who are considered the most privileged from the cognitive and social points of view. In fact, it seems to us much more contemporary than those types of narratives that have long since disappeared from the research scene, except as anecdotes (☞ **Fabricating Science**).

Thompson's book is crawling with them. For instance, he also recounts how an orangutan at a zoo stole an orange, while the keeper pretended to sleep so as to spy on him, and hid the peel to erase any trace of his misdeed. This scenario evokes for us, in a clear way, two different species of beings exercising the art of lure and deception. And yet, this is not what Thompson saw. Astonishingly, he rarely used these terms, nor those of lying or trickery. He saw something else, and his interpretation was guided by the problem that he was attempting to solve: to create a sense of common intelligence between animals and humans so as to better protect animals. It is rather difficult to constitute such commonality within the context of a creationist anthropology, which assumes a universe ruled and hierarchized by multiple divine decrees that prohibit animals from having a soul. With good insight, Thompson therefore attempts to construct this commonality by basing it on a series of analogies of similar inductive intelligences and sensitivities.

The case of the stolen orange will be taken up again a few years later by Darwin. But with Darwin, it is no longer the lying that comes to qualify the act. It's the shame. If the ape hides the orange, Darwin says, it's because he is conscious of prohibition; one might assume that with this behavior, which is so similar to that of children, we are dealing with a precursor of moral sentiment. The same story, a different interpretation: in subjecting it to a regime of the line of descent, Darwin's project is no longer one of constructing proximity but a regime of continuity. Behavioral analogies constituted the most promising indicators, especially if they had to do with a key domain of human exceptionalism: morality.

It is remarkable that these kinds of animals, and the narratives that they inspired, will completely disappear from the scientific scene. Too anecdotal and too anthropomorphic, these stories have been reduced to the knowledge of amateurs who do not forbid themselves from continuing to cultivate or marvel at them. Zoos will thus become one of the main sanctuaries of lies, especially in the rotten tricks that certain animals play with their keepers to escape or break up the boredom.

It will take almost a century before scientists consider resuming the question by explicitly tying it to mental states. Examples of intentional lying and trickery will begin to proliferate in the field by the beginning of the 1970s, and less than ten years later, they'd enter the laboratories. At Gombe Stream in Tanzania, Jane Goodall observed chimpanzees who would feed themselves on bunches of bananas that she had left out for them. A young chimpanzee came along and was about to help himself when a dominant male saw him in the area. The behavior of the young chimp changed immediately: he took up a detached air, as though entirely indifferent to the bananas. The older male headed off, and with the coast now clear, the younger one returned toward the bananas, when suddenly the older male reappeared. Suspicious of the apparent casualness of the young one, the older one had hidden to observe him. Other events from the field will come to confirm what Goodall's observations imply: that chimpanzees are liars.

It is only at the very end of the 1970s that these observations will obtain a significance that will trigger an impressive stream of research, such that they'll pass from anecdotal status to real scientific projects. Note that within this context, the expression "obtains a significance" recovers a very specific sense, as it highlights the fact that anecdotes have become "significant" because they've passed experimental tests: they've been demonstrated in the laboratory. From out of this they've acquired the status of a legitimate research subject. Controversial, but legitimate. After working for several years with chimpanzees, David Premack and George Woodruff decided in 1978 to give a new direction to their work.[2] Up until this point, they explain, scientists had been testing the chimpanzees as "physicists," as chimpanzees were generally asked to solve problems like catching a banana with a stick, stool, and crate. Moving forward, they would now test "psychologist" chimpanzees. Are apes mentalists? Are they capable, as is commonly said, of reading the minds of others? Can they, in other words, put themselves mentally in the place of others and attribute to them intentions, beliefs, and desires?

According to the two researchers, the experiment would be conclusive. If the experimenter were looking for a candy, and the chimpanzee knew where it was hidden, the latter would generally help if he knew that the human would give it to him. However, if the human kept it for himself, it would be observed that the animal would lie to him in the next attempt. This shows that, on one hand, the chimpanzee grasped the fact that the human had intentions and, on the other hand, that what the chimpanzee knew about the situation didn't correspond with what the human knew. The chimpanzee therefore perceived that what the human had in mind was different from what he himself knew.

Of course, in anticipation of the rather predictable reaction of behaviorists—if it's not conditioned, there's no salvation—and their famous Morgan's canon (☞ **Beasts**), the two authors concede that one can always reduce the explanation to the much more simple hypothesis of conditioning: the chimpanzees were doing no more than obeying the so-called rules of learned associations. So the chimpanzees have not actually been able to work out the intentions of the one who betrayed them; they are only associating, mechanically (due to being confronted), the absence of compensation with the researcher responsible for this. Once bitten, twice shy. If this is the most elementary faculty for learning by conditioning, it doesn't require any special competence. Premack and Woodruff will thwart this argument, and not without humor, by inverting the hierarchy of abilities and turning Morgan's canon against those who tend to invoke it: we spontaneously attribute intentions to others because it's the simplest and most natural explanation, and, they say, the ape probably does the same: "The ape could only be a mentalist. Unless we are badly mistaken, he is not intelligent enough to be a behaviorist."[3] This leads one to wonder if chimpanzees in fact have less difficulty attributing mental states to other species than do the behaviorists.

Leaving the cognitive laboratory behind, the capacity to lie returns to the field to help rescue a new definition of the social chimpanzee that has been proposed. Following upon the discoveries of horrific fights, crimes, and cannibalism, the chimpanzee was relieved of his role as a peaceful noble savage, which he had previously held; but now his skills at lying will give him a new role, as he becomes the "Machiavellian chimpanzee" endowed with an essential political quality: the power to influence, that is, to manipulate, others.

Other animals, in their own turn, will lay claim to this skill. Chimpanzees, naturally, will lose their monopoly to other apes and monkeys. In light of the privilege accorded to the neocortex for the evolution of this faculty, birds a priori do not appear to be good candidates (☞ **Magpies**). However, the great sociality of ravens, together with an observation in the field, led Bernd Heinrich, a raven specialist, to rethink this prejudice.[4] What Thompson's ape did to a representative of the crow species in severely plucking it after he had been misled here finds its "corvidean" counterpart, this time with a swan as the victim.[5] The swan was in the process of incubating some eggs when a couple of ravens tried in vain to steal them by attacking the swan. The swan was threatened but didn't move. One of the two partners then did something never seen before among ravens: he pretended to be injured (what is known with other birds as feigning a broken wing). And once again, the swan went in pursuit of the pseudo-wounded . . . while the other raven rushed over to the nest and took an egg. Feigning an injury has nothing remarkable about it, of course, because many birds who nest on the ground do this when they need to attract a predator away from the nest where their chicks are, by simulating the act of being wounded and pretending to run away with great difficulty, thereby leading the danger after them. But this behavior had until this point been called a preprogrammed mechanism, so it had never required another explanation as such; selection sufficed as a reason, and in doing so counteracted the possibility of attributing mental states. Is it because the ravens exhibited this behavior in a new context and in an entirely nonhabitual way that the explanation of "instinct" was not used? Or was it because Heinrich was confident enough in the intelligence of these ravens that he could offer them a less reductive version? It is difficult to determine, and it probably should not be done. But if this question makes us hesitate, it's because it is probably time to reopen it for those who had thought it had been answered.

If the example of the raven is convincing for those who have observed it, then for the experimentalist, within the context of the intense rivalry between field research and laboratory experimentation, it still falls within the category of an anecdote. Rare events, as the name suggests, have little chance of being repeated—unless, of course, one can imagine a dispositive that requires animals to demonstrate their reliability. Heinrich did this with ravens held in captivity, and a number of experiments will confirm the accuracy of his hypothesis. If a raven senses that he is being observed by a conspecific, he will make a show of hiding his food in one spot but will really conceal it elsewhere while the other raven looks for it in the supposed hiding spot. Just as experimenters do with their apes, Heinrich also conducted some experiments that involved the researchers.[6] The latter would rely on a practice that is rather frequent among ravens in captivity: they like to play by hiding objects. And yet, if a human observer steals one of the toys that has been hidden within the context of play, one notices a radical change of attitude with respect to this same person once food is involved. The raven will assume many more precautions, such as ensuring that he is out of the observer's line of sight and devoting more time in recovering the hidden object, than he would if he were in the presence of someone unknown. This means, therefore, that not only are they conscious of the intentions of their conspecifics but they can expand the circle of those to whom they attribute intentionality, including humans, within this game of sociality.

Pigs have also come to be included in this large family of liars. An experiment in a maze brought together a pig "informed" with the hiding spot of some food and an "uninformed" pig: it showed that if the "uninformed" one profited from the other's help in digging up and eating the hidden food, then in the next test, the informed pig would casually misdirect him into a dead end of the maze.

In addition, the possibility that an animal might attribute mental states or intentions to another will enhance new alliances between relatively divided areas of research: those of the cognitivists, who work mainly in laboratories in conditions that sometimes resemble a school examination, and those of field primatologists, who are more preoccupied with the sociality of their animals. This alliance takes the form of a hypothesis: because lying is based on the possibility of understanding the intentions of others, it should correlate with social cooperation. Altruism and deception are two sides of the same aptitude, social subtlety. The world demoralizes and remoralizes itself, and the researchers, who used to be rivals, now cooperate.

Other concerns will also play a role for the interest in these dishonest animals, as well as confirm the rise of this subject of research. For example, sociobiologists are interested in the way that animals use trickery to resolve conflicts of interest. How does one resolve, for instance, a conflict when it occurs between two potential future parents when each must ensure that its partner will take care of the nest? According to sociobiologists, each must ensure that it invests the least possible while making sure that its partner doesn't do the same. Lying propaganda and shameless manipulations become the rules of etiquette [*savoir-vivre*], in the most literal sense.[7] The dunnock, a bird from our region of Europe, has invented a pretty amazing system. And just this once, this invention is credited to females, more specifically only to some of them, because they do not all exhibit this behavior. In certain circumstances, a female whose territory borders on that of two males will act in such a way as to convince them both that they could be the father of the brood she laid. According to observers, if she does this convincingly, she will find herself with two males to defend a larger territory and feed the young. Her strategy consists in mating, with the greatest discretion possible, with one and then the other. They discover the truth sooner or later, but neither can be absolutely sure that he is not the biological father. And as the breeding season is well under way, they can no longer reverse their steps and take the risk of an uncertain defection.

We are definitely dealing with the typical and fairly recurrent schemes of sociobiology: conflicts of interest between males and females, animals manically mobilized for reproductive problems, dilemmas between short- and long-term investments, carefully calculated reproductive capital whose evaluative strategies would make even the most cynical traders tremble. To be sure, the scenarios change when they feature females who are subject or victim to the dominance or fickleness of males, but this hardly changes the image of nature as subject to laws of competition. Cooperation, do not forget, is no more than the result of a dark conspiracy.

One last interesting hypothesis nevertheless emerged from the overlapping research of cognitivism and sociobiology. The act of lying and the need to protect oneself from deception should have, from an evolutionary point of view, led toward an arms race—one can see again the privileged schema of sociobiologists. In a world of liars, the problem is how to develop a double ability: on one hand, protecting oneself from liars and learning to detect trickery and, on the other, becoming a good liar oneself. According to the model of the arms race, the more lying develops as a skill, the more the skill of lie detecting ought to evolve as well; following from this, lying should tend to become more and more imperceptible and the detection of it even more subtle. This ability to lie in an unexpected way should have therefore led to its extreme and produced a strange capacity of delusion: that of lying to oneself. In other words, in a world of lie detectors, nothing is as efficient in tricking others as believing in one's own inventions oneself, thus becoming the deliberate victim of unconscious motivations.

As one can see, lying borrows from the most heterogeneous domains and brings together various cognitive and disciplinary types, modes of psychology that the sciences have carefully separated and to which it in part owes its success: it arises out of biology; it involves sophisticated cognitive modes, beliefs, and mental states that are of interest to cognitivists and, it's worth noting, analytical philosophers; it's now creating ties with unconscious processes; it supports sociological and political theories; and above all, it is considered closely articulated with areas of morality: lying, empathy, understanding the desires of others, and care for the other all co-emerge.

Upon reading of this surprising alliance, a final remark is needed to conclude this piece of the story that has placed animals on the path of counterfeiting. In these research studies, and as can be found in many theories of evolution, there is a paradox that is not lacking in humor: at one point or another, the behaviors that for us are the most clearly stigmatized by the moral among us come to resemble the noblest of virtues as soon as they are retranslated by the theories of natural history and evolution—or, at the very least, are the condition for it. To put it another way, what an animal does, and what morality finds repugnant and condemns unequivocally, becomes, in the context of nature, the most certain path toward morality. Male jealousy now stabilizes couples, the most inflexible and arbitrary hierarchy secures social peace, and lying proves to be, still from this perspective, evidence of the highest consideration for another and the basis for cooperation. One wonders sometimes whether ethology wasn't in fact invented by some Jesuit who is fond of mischievous casuistry. As opposed to a hell that is paved with you know what, another image could instead be posed: that of a paradise where, eventually and most likely, the worst of intentions lead.

Q
FOR QUEER

Are penguins coming out of the closet?

Queer: strange; odd. Slightly ill. *Usage*: the word *queer* was first used to mean "homosexual" in the early 20th century. . . . In recent years, however, many gay people have taken the word queer and deliberately used it in place of *gay* or *homosexual,* in an attempt, by using the word positively, to deprive it of its negative power.

—*New Oxford American Dictionary*

Between 1915 and 1930, a group of penguins lived at the Edinburgh Zoo. Over the course of these years, a troop of zoologists meticulously and patiently observed them, beginning by naming each and every one of them. But first, before receiving their names, each of the penguins was placed within a sexual category: on the basis of a couple, some were called Andrew, Charles, Eric, and so on, whereas others were christened Bertha, Ann, Caroline, and so forth.

As the years passed, however, and the observations accumulated, more and more troubling facts seemed likely to sow disorder within this beautiful story. To begin with, one had to face the facts, as the categorizations were based on a rather simplistic assumption: certain couples were not formed by a male penguin and a female "penguine," but from among all penguins. The permutations of identity—on the part of the human observers, not the birds—had a "Shakespearean complexity" to them. In addition to this, the penguins themselves decided to put their own stamp on things and make things even more complicated by changing their couplings. After seven years of peaceful observations, it was therefore realized that *all* but one of the attributions were wrong! A complete overhaul of the names was thus carried out: Andrew was rechristened as Ann, Bertha turned into Bertrand, Caroline became Charles, Eric metamorphosed into Erica, and Dora remained Dora. Eric and Dora, who spent their days peacefully together, were now called Erica and Dora, whereas Bertha and Caroline, who were known for some time to be homosexual, were from then on known as Bertrand and Charles.[1]

These observations, however, were not going to damage the image of nature. Homosexuality remained a rare phenomenon in the animal world, and these penguins were probably just a few pathological cases observed here and there on farms and in zoos and thus assumed to be due to conditions of captivity—which fell into perfect agreement with human psychopathological theories that equated homosexuality with mental sickness. Homosexuality was definitely unnatural, as nature could testify. But it seems that, in the 1980s, nature had a change of heart. Homosexual behaviors were now everywhere. One was probably supposed to imagine, during these same years, disastrous consequences from the queer revolution and the American gay movements that would contaminate innocent creatures.

But the question should no doubt be posed differently: why hadn't homosexuality been seen in nature until this point? In the book *Biological Exuberance,* Bruce Bagemihl considers a number of hypotheses following his long investigation reporting on species that had recently come out of the closet. To begin with, he says, homosexuality wasn't seen because nobody expected to see it. There wasn't a single theory available to meet the facts. Homosexual behavior appeared to be a paradox of evolution because, in principle, homosexual animals did not transmit their genetic heritage. This stems in fact from a very narrow conception of sexuality, on one hand, and of homosexuality, on the other. For the former, animals only mate due to the goal of reproduction. The strictest god would have succeeded in obtaining from animals a virtue that he had not been able to with any of his faithful humans. Animals don't do a thing except if it is useful for their survival and reproduction (☞ **Necessity;** ☞ **Oeuvres**). For the latter, homosexual animals would be exclusively oriented toward partners of the same sex and, in this respect, would be proof of a strict orthodoxy.

Next, for those who observed behaviors oriented toward a partner of the same sex, a functionalist explanation could justify them perfectly well, and it had the merit of removing this behavior from the sphere of sexuality. When I was a student, we learned in an ethology course that when an ape presents his or her genitalia to another and allows himself or herself to be "mounted"—I also heard this said of cows—it has nothing sexual about it; it is just a way of affirming dominance or submission, depending on the position adopted.

Lastly, another reason that has had considerable influence is the fact that researchers have only observed but a few homosexual behaviors in nature because they are so rarely seen. Not that they are rare, but that we don't see them. Just like we rarely observe heterosexual behavior, because animals, who are very vulnerable in these moments, generally do it in hiding so as not to be seen, especially by humans, who are seen as potential predators. And since we see newborns emerge every year, nobody has ever doubted that animals have a sexuality, even if it is only seen on rare occasions. But rare does not mean "not at all," and this concerns homosexual behavior as well. How is it that this has remained unmentioned for so long in research studies? The primatologist Linda Wolfe queried her colleagues on this subject at the end of the 1980s.[2] On the condition of remaining anonymous, many of them admitted that they had seen such behavior, with males just as much as with females, but they were afraid of homophobic reactions and of being seen as homosexuals themselves.

In light of these reasons, therefore, one can legitimately think that the queer revolution changed things. It opened the idea that forms of conduct that were not strictly speaking heterosexual could exist, and it encouraged researchers to look for them and speak about them. Hundreds of species now participate in this revolution, from dolphins to baboons, as well as macaques, Tasmanian geese, Mexican jays, gulls, insects, and, of course, the famous bonobos.

At the same time, animal sexuality benefited from what I would call their "cultural revolution." After having been excluded, animals can now claim to be within the order of culture. They have artisanal traditions (for tools or weapons); fashionable songs (with whales, for instance); practices of hunting, ways of eating, medications, and dialects that are specific to groups from now on christened as "cultural"; and practices that are acquired, transmitted, abandoned, or undergo waves of invention and reinvention. As such, sexuality is now a candidate, and this includes its homosexual dimensions. It also carries the mark of cultural acquisition.

The ways that acts are performed—for example, among female Japanese macaques—demonstrate these differences: some practices appear to be more popular within some troops, and they evolve over time, with some inventions tending to supplant other ways of doing things. Some "traditions," or models of sexual activity, can be invented and transmitted across a network of social interactions, moving between and within groups and populations, geographies and generations. According to Bagemihl, sexual innovations in a nonreproductive context have contributed to the development of other significant events from the point of view of cultural evolution, most notably in the development of communication and language as well as in the creation of taboos and social rituals. Among bonobos, twenty-five sign-language signs have been found to indicate an invitation, a desired position, and so on. These signs can be transparent and their meaning immediately decipherable, but some of them are more codified and require that the partner already know them to understand them. The gesture of inviting a partner to return, for example, is in one group executed by making one's hand turn in toward itself. Their opacity and stylization invite one to think that there are abstract symbols here. The order of the gestures, which is equally important, leads to the hypothesis that animals may be able to use syntax. In terms of the organization of relationships, they seem to be marked by complex codes. According to Bagemihl, rules guiding how to avoid others are, with certain species, relatively different if it consists of hetero- versus homosexual relations; what seems to be not permitted with one sort of partner might be permitted with another.

To focus on the diversity of these practices, as Bagemihl does, is an explicitly political issue, and one with many positions. On one hand, this diversity takes sexuality out of the natural domain so as to situate it within a cultural one. It's an important issue and one that constitutes a choice. It is not just a case of removing homosexuality from the sphere of mental pathologies or from legal domains—in some U.S. states, it still continues, as we will see. Bagemihl will refuse the hand stretched out to him, the allies who could have strategically helped to depathologize and decriminalize homosexuality. In the outstretched hand there is this simple proposition: if homosexuality is natural, it is therefore neither pathological nor criminal. The argument for its unnaturalness has also been used during a trial by a judge from Georgia—in the *Bowers v. Hardwick* case. Caught in the act of homosexual relations, Hardwick was sentenced, and the unnaturalness of the act was used among the arguments justifying the accusation. Naturalizing homosexuality could take care of a lot of things. For Bagemihl, even if homosexuality is natural, it cannot be figured into the equation "what is natural is right." Nature does not tell us what *ought* to be from what is. It can feed our imaginations but not compel our actions. It is worth noting, in passing, the irony of this story. Despite this refusal, Bagemihl's book will be invoked, in 2003, during a trial that featured the Texas court system versus two homosexuals, Lawrence and his partner, who were caught in bed together by the police following a report of a nighttime disturbance of the peace. On the grounds of the previously mentioned judgment, that of the *Bowers v. Hardwick* case, they were prosecuted for homosexuality. The Texas judges, however, refused to follow the case law set by the precedent judgment and refuted, on the basis of Bagemihl's book, among other reasons, the argument of naturalness.[3] At the end of the trial, the antisodomy law was considered to be anticonstitutional.[4]

The author of *Biological Exuberance* had another, less theoretical reason for refusing to record homosexuality as a fact of nature. Bagemihl is not only homosexual. He is queer. To quote him, what interests him is "the world as 'incorrigibly plural.' . . . It suffers difference, honoring the 'anomalous' and the 'irregular' without reducing them to something familiar or 'manageable.'"[5] The meaning of being queer cannot be better defined. It is a political will. And this political will does not only concern humans. It concerns the world around us. It concerns our ways of entering into relations with this world and, among these relations, of knowing and practicing this knowledge. Bagemihl measures the risks of accepting whether homosexuality is natural. It seems to be the object of biologists to try to resolve this paradox, and he knows very well which biologists are already on the case: it's the sociobiologists. They have, in effect, buckled down with an insatiable appetite for this new problem: it's another case that will come to illustrate and expand their theory. It will be even more "all-terrain" [*tout-terrain*]; the world will be sociobiologized. For the theory of kinship has a solution entirely found in homosexuality, though it rests on a strict conception of an orthodox homosexuality. Of course, homosexuals do not transmit their genes to their descendants, so normally they ought to disappear because of a lack of descendants carrying this gene—it goes without saying that homosexuality is genetic. Homosexuals, however, direct their attentions and their abundant leisure time (because they don't have any dependents) toward their nephews, who are carriers of an identical part of the genetic heritage. It is therefore through these latter descendants that the gene continues to assure its propagation. This type of biology is political, not only in the sense for which we usually reproach it—these theories can easily be retranslated into misogynistic, racist, eugenic, capitalistic, and so on, theories—but in the sense that, to put it simply, these theories animalize, insult, and impoverish those for whom they pretend to take account. In other words, sociobiological theory—to recall the words of the psychologist Françoise Sironi—is an abusive theory. Every behavior is reduced to a genetic purée; beings become blind imbeciles determined by laws that escape them—and that prove to be disturbingly simple. No more inventions, no more diversity, no more imagination—and yet, if they still persist, it's because they have been selected to allow us to spread our genes. One cannot be both queer and a sociobiologist.

But can we really say that animals are "truly" homosexual, in the same sense as we can be? Bagemihl responds: but can we say this even of ourselves? Can we name, under the same term, the same realities, from the amorous youths of ancient Greece to the most diverse modes of being today? And can we say that, among animals, the entire range of forms of relations that organize between the same sex are "truly" the same (☞ **Versions**)?

It's here that I find the coherence of Bagemihl's project. Biology must respond to the diversity and exuberance of nature and beings; it must rise to the level of what is required of it. This reflects the bias of what he says about the scientific task: multiply the facts to allow a chance at multiplying interpretations. This is far from the "all-terrain" theories; the diversity of things will fertilize the diversity of interpretations. This is what he elsewhere calls "doing justice to the facts."

Nature is invited to a political project. A queer project. It teaches us nothing about who we are or what we ought to do. But it can feed our imagination and open our appetites for the plurality of usages and modes of being and existing. It never stops recombining categories and re-creating, from the multidimensionality of each and every one of them, new modes of identity. What is meant by being male or female, for example, can be found among many animals according to inventive modes that are similar to a multiplicity of ways of inhabiting a gender. One can find among certain birds—and sometimes even among members of the same species—two characteristic situations: on one hand, one can find females living an entire life as a couple, making a nest together each year, incubating eggs that one of the females has fertilized in mating with a male, manifesting regular courtship behavior toward one another, and yet never showing any mating behavior. On the other hand, one can find a male mating all his life with the same female, with whom he mates regularly and raises the young, but who, on occasion, mates with a male (and never does so again). How do you categorize this? Are these relations homosexual? Bisexual? Are these birds consistently female or male? Are these even good categories to take account of what they're doing and who they are?

I recognize here a project that I was able to find in the writings of Sironi, based on her work with transsexual and transgendered people. The queer project that she supports roots itself within questions of sexual and gender identity, but its political aim is first of all tied to a practice that obliges us to think and that calls for thought. These two approaches, however, aim to transform habits, transform relations to norms, to oneself and others, and to open possibilities. So if this clinician's will is to learn, along with those who address themselves to her, how to help them fight against the "abusive theory" that her colleagues exercise against them, to "free gender from its normative shackles," and to support "its amazing creative vitality," she relies just as much on them—those who are the experts of metamorphosis—to help us to think and imagine different "contemporary identity constructions."[6] "Transidentitary and transgender subjects have a function, currently, in the modern world. . . . Their function is to enable becomings, to show diverse expressions of multiplicity in itself and in the world":[7] to deterritorialize oneself, to open oneself up to new *agencements* of desire, to cultivate an appetite for metamorphoses, and to forge multiple affiliations.

R

FOR REACTION

Do goats agree with statistics?

"In most research," wrote Daniel Hestep and Suzanne Hetts in 1992, "the scientist aspires to have the animal behave toward the investigator as if he or she were a socially insignificant part of the environment. This reduces communication between the two to a minimum. Many field workers and some laboratory investigators go to great lengths to either conceal themselves from their subjects with blinds or use remote sensing devices (binoculars, radiotelemetry devices, etc.) to accomplish this goal. Others spend enormous amounts of time and energy habituating the animal to the presence of the investigator. How well these attempts succeed in reducing the reactivity of the animal to the researcher is difficult to assess and is rarely addressed directly. Investigators do not often describe how their subjects react to them."[1]

The authors are right, there aren't many counterexamples. A few can be found among primatologists (☞ **Corporeal**), or even with Lorenz, who, as it happens, used the intimate relations he wove with his animals to study them. The fact that for most of them this didn't happen without trouble reflects the difficulty. Things are, in this respect, progressively changing, and the critique that can be read between the lines is proof of this new attitude concerning animal observation. Even if I agree with this critique, there are still some things in its formulation that deserve going back over. The essay from which this quotation is taken can be found within a collection of studies that has united scientists wanting to reflect on and explain the relations that are created between an animal and his or her observer. The project is exciting. This essay shows, however, the limits that still remain within this type of attempt: the authors speak of "reactions" and "reactivity." With Haraway, I have learned to pay attention to terms, not only because they translate habits, but above all because they engage

139

in narratives that are far from innocent (☞ **Versions**; ☞ **Necessity**).

The term *reaction,* a familiar one among ethologists, is not without consequences. It remains, within the context of research on relations, off track with what it purports to explore. On one hand, in reducing how an animal accounts for an observer's presence to a "reaction," the authors perpetuate the notion of a passive animal, entirely determined by causes that are beyond him and over which he has no control. On the other hand, and this is related, by considering habituation as a method designed to diminish the "reactivity" of animals to the presence of an observer, it overlooks the fact that animals take an active role—a very active role— in the encounter. This decrease of reactivity is in reality nothing but the effect of something entirely else; it doesn't explain anything but instead asks to be explained. For each troop, a number of hypotheses still need to be considered that are not only contextualized but also dependent on how groups organize, how they interpret the intruder, the opportunities that he may offer, and so on. In short, every ethologist finds herself in a position similar to that of an anthropologist when she poses (or attempts to answer) the inevitable question of field research: how do those whom I am encountering understand what I am trying to do? What intentions do they attribute to me? How do they translate what I am searching for? How do they assess what I bring as a problem or as a benefit, and for whom? As soon as primatologists—or, more rarely, ethologists—pose these types of questions, another story begins to impose itself. It is in this way, then, that on the basis of a routine observation, the primatologist Thelma Rowell proposed to revisit what is understood by the term *habituation.*[2] When compared with ape troops that have simply been surveyed now and then (or observed from a distance), there are demographic changes in certain troops that had benefited from the presence of an observer who practiced habituation. The term *benefited* is not chosen by accident, for the demographic changes would be more favorable for the latter. In paying attention to the conditions in which the process of habituation is shaped, Rowell realized that the close presence of a scientist discouraged predators who were forced to hunt elsewhere. This led her to hypothesize that a number of animals deliberately allowed the observer to approach when they understood that their presence protected them. It is therefore not about habituating but rather composing with, or even utilizing, the observer. But this explanation is not generalizable. Some apes do not have any real problems with predators, others only have serious trouble with humans, and still others, like the less social orangutans, must learn to compose themselves with the intruder, who makes their conspecifics and females flee. Reactivity is not even in the picture here. We have an entirely different story—and an entirely different way of making stories—taking shape. Now it requires that the beings experience the encounter, that they interpret it one way or another in terms of what is at stake, what is at play in the exchanges, and that they subtly negotiate them. Obviously, this goes against the requirements of "doing science" that many researchers must abide by (☞ **Laboratory**).

To renounce reactivity and to do so seriously—that is, to draw the consequences that this decision requires—is not easy for a researcher. It's a difficult decision. It often means seeing one's work disqualified and articles refused. To renounce reactivity entails considering that animals actively take into account and respond to a proposition that is made to them, which commits the researcher in another way. For if *response* implies a possible bifurcation, *reaction*, by contrast, entails that the way that the problem has been posed overdetermines what will happen next and the meaning that it has. This means that, for the researcher who agrees to listen to the animals who respond, the control over the situation is distributed differently. If I were to put this the way that Stengers translates this difference, I would say that the scientist will be *obliged* by the response, that he or she will have to respond to and respond for them.

The researcher Michel Meuret has made this choice: he allowed himself to be guided because the animals that he was observing were responding to him, which, as a last resort, compromised any possibility of sampling, including the consequences this could have for the possibility of being published.[3] His situation is all the more interesting because it was relatively unexpected. It consisted, of course, of a practice of habituation, but it was carried out in a certain experimental framework. The animals were not apes but rather goats. Even more surprising, Meuret was not studying social behaviors but food preferences, a subject that doesn't often persuade researchers to pay sustained attention to animal sociality.

His research project aims to evaluate what goats eat (what exactly do they eat, in what quantity, and how) in unusual conditions—in this case, areas that have been cleared of brushwood. It's true that the entire dispositive rather resembles an investigation similar to that of field ethologists, but the "unusual conditions," which is to say food that is not used within animal farms, justifies the term "experimentation": the goats are proposed a "test," and they are evaluated on the way that they respond. The experiment begins with a first step of *reciprocal familiarization* between the observed animals and their observers. Once the familiarization seems to have been achieved, the researchers, with the help of a herder, will attempt to identify the animals that can be followed and that they anticipate to be less bothered by the permanent presence of an observer. The research begins once this step has been accomplished. From this point on, each member of the research team follows a chosen animal every day and observes what he eats throughout the day. Each detail is carefully noted, every species of plant inventoried, every bite recorded. The proximity is total, the interest in the observed sustained.

The scientific method demands that animals be chosen indiscriminately to constitute a random sample. Now, as it happens, and this is why there is a second step, this decision can leave nothing to chance. This could prove to be disastrous. The continuous presence of the observer could, for example, contribute to a change in the social status of an individual. An aspiring leader could translate a researcher's interest as encouragement. The fact of being the object of intense interest by a human leads some goats to act like they want to supplant others, take their food, and even pick fights with them. For others, being the object of human attention provokes aggression from their companions, as if this interest translated a willingness of the goat to change her place in the hierarchy. The risk is not only of creating disturbance in the group but also of no longer knowing what one is observing: is this what a goat eats in unusual conditions or, on the contrary, is this what a goat eats when he wants to show to others his superiority, because all of a sudden he thinks his status has changed?

The percentage of goats that can be followed is no more than 15 to 20 percent of the herd. This is not a sample. This means that the observed animals are in no way representative of the herd, and even less so of goats in general. But they can nevertheless be proof of something else with respect to goats, namely, their approval or disapproval of the quality of what is offered to them in these unusual areas. They could therefore be thought of not as (statistically) representative of goats but instead as *representatives* (or "delegates") to the researchers and people who want goats to look after the maintenance of areas that have been cleared of brushwood, which is imperative in regions susceptible to forest fires.[4] And they will be *reliable representatives* if the scientists have selected them correctly. This terminology, though it isn't explicit in this dispositive, takes good account of the practice and relations established. Even if it is implicit, it makes generalizations a lot more hesitant and researchers much more attentive to the consequences of their decisions, their work, and the ways that goats respond to them. If, in the course of observation, a goat shows too much interest, anxiety, or discomfort because of the close and constant presence of the observer, Meuret explains, the observer must discontinue his observations. To be a representative [*représentante*] is to be the one who guarantees the reliability of the dispositive and the strength of the results; this implies neither indifference nor reactivity to the practice of observation but an endorsement or approbation [*probare*] that counts as proof. On the part of the researchers, this means imagining that their animals respond to and judge their propositions and that the researchers receive a response in return for the judgment. As Meuret said to me, "a good sign at the start of an observation is when an animal pushes you because you are in the way of what he wants. This means that he is capable of showing that you are bugging him."

Some experimental research studies are beginning to consider this idea that it is much more interesting to address oneself to a reliable representative than it is to address a barely interested representative. They are rare. The successful studies with talking animals are examples of this (☞ **Laboratory**). The animals who do not want to speak will not collaborate anyway. Researchers are thus only obliged to work with those who show interest and to actively solicit them in this sense: that they become interesting. Other initiatives of this kind emerge, however. Quite recently I discovered that some primatologists of the Yerkes National Primate Research Center in the United States carried out an experiment with captive chimpanzees to evaluate the influence of personality on the act of imitation with tool usage. If two chimpanzees with quite different personalities—one who is young and one who is older and more dominant—both show to their conspecifics how to manipulate a tool to acquire candy, which of the two chimpanzees will the spectators tend to imitate? The two manipulations being taught are slightly different, which allows the privileged one to be identified. This research, I should note in passing, is designed to understand the mechanisms of the cultural diffusion of a new habit: the young are generally the ones who invent, whereas the dominant ones often have more prestige. It seems, at least in matters of experimental tools, that prestige prevails, which simply leaves the paradox intact: it is still not known how an innovation is transmitted. But this is not what I wanted to highlight with this research, nor is it reported in the article, except in the methodological section, as is often the case. The researchers write, "The chimpanzees recognize their names and can be 'asked' to participate in studies by calling them inside from the outside enclosures, or placing [an] apparatus at the enclosure fence and giving them the choice to interact with it."[5]

It is only a small step. But it might promise others. Of course, that the chimpanzees were recruited in conditions that required their interest does not indicate that these types of questions interest them; the fact that the notion of dominance is still at the heart of researchers' preoccupations would tend to make me hesitate (☞ **Hierarchies**). But when this small step is taken by someone like Meuret, it makes me think that a certain conception of objectivity is replacing the one that defines knowledge as an act of power so powerful that it claims to be a view from nowhere. Objectivity is no longer, as Haraway suggests, a matter of disengagement but about "mutual *and* usually unequal structurating."[6] This new way of conceiving of objectivity requires, she writes, "that the object of knowledge be pictured as an actor and agent, not a screen or a ground or a resource. . . . The point is paradigmatically clear in critical approaches to the social and human sciences, where the agency of people studied itself transforms the entire project of producing social theory. Indeed, coming to terms with the agency of the 'objects' studied is the only way to avoid gross error and false knowledge of many kinds in these sciences. But the same point must apply to the other knowledge of many kinds in these sciences. . . . Actors come in many and wonderful forms. Accounts of a 'real' world do not, then, depend on a logic of 'discovery,' but on a power-charged social relation of 'conversation.' The world neither speaks itself nor disappears in favour of a master decoder."[7] I'll let her conclude: "Acknowledging the agency of the world in knowledge makes room for some unsettling possibilities, including a sense of the world's independent sense of humour."[8]

S

FOR SEPARATIONS

Can animals be broken down?

"While studying wild baboons in Kenya," Barbara Smuts writes, "I once stumbled upon an infant baboon huddled in the corner of a cage at the local research station. A colleague had rescued him after his mother was strangled by a poacher's snare. Although he was kept in a warm, dry spot and fed milk from an eyedropper, within a few hours his eyes had glazed over; he was cold to the touch and seemed barely alive. We concluded he was beyond help. Reluctant to let him die alone, I took his tiny body to bed with me. A few hours later I was awakened by a bright-eyed infant bouncing on my stomach. My colleague pronounced a miracle. 'No,' Harry Harlow would have said, 'he just needed a little contact comfort.'"[1]

I can't hold it against Smuts that she made reference to Harlow; the reference is in fact unavoidable because it is found in her review of a book devoted to his biography, written by the journalist Deborah Blum in 2003.[2] Nevertheless, if I raise the possibility of a complaint, it's because, in the present day, it is still almost impossible to speak about attachment, even among humans, without evoking his name. As though it is due to him that we know that when an infant is separated from any meaningful contact, psychological and/or physical death follows. We knew this already! To give him credit that we know this is to implicitly endorse the manner by which he proposed to "know" it: through a system of evidence, which, in this context, means a system of destruction. It is now time to speak of him as a historical event, as "something that happened to us" and that obliges us to think.

To evoke Harlow, as one of Smuts's colleagues has, by claiming that he "would have said" is not thinking in a serious way but rather stating that we have learned nothing, all the while pretending to know. For Harlow would not have "said"; he would have done. If Harlow had been there, we

would have had an entirely different story. The psychologist would have
inevitably found the chance to test some more, on one more species, the
thesis that he claimed to have proven. He could have once more tinkered
with the wire mannequin and the terry cloth mannequin and verified—
once again, one more time—in the test imposed on the orphaned baby
baboons, that attachment is necessary. Ultimately, Smuts's colleague had
good reason to pronounce this a miracle. For a miracle it was. Not, however, in terms of the unexpected return to life of this young orphaned
baboon but in terms of what could make a scientist forget that one never
knows better those one is questioning than when one accepts to learn *with*
them, and not *about* them, which is to say *against* them. Smuts learned—by
listening to what spoke to her of compassion, by subjecting herself to the
risks of attachment—in just one night what years of torture had entitled
Harlow to produce as knowledge. Smuts learned what she already knew,
but that we never stop relearning each time we are touched: that one can
only truly understand others, above all in these stories of attachment,
by allowing oneself to go through, with these very attachments, what is
important to them.

"We have spent the last 4 yr deep in depression—fortunately, the depression of others and not ourselves—and we regard this period of animal
research as one of the most succinct and successful we have experienced."[3]
This is how, a few years later, he described the results of his research. And
yet, he clarified that it was not really depression but rather love that was
at the heart of his preoccupations: "Oddly enough, we initially produced
monkey depression not through the study of lamentation, but through
the study of love."[4] How does one go from depression to love, or from
love to depression? Harlow's research is famous today; by studying the
consequences of the absence of relations on the development of baby
macaques, the psychologist was endeavoring to prove and to measure
relations' vital importance.

It's worth pausing on what, in a psychological laboratory, could be
meant by "studying love." Blum's biography of Harlow, despite its un-
easiness and obvious ambivalence (her previous book didn't hide her
sympathies with protectionist and activist movements),[5] brings to the
fore what I would call the poison of this legacy: she makes Harlow into a
revolutionary hero who obliged the psychological world to accept affection

as an entirely legitimate subject of research. And she reconstructs his career by finding signs that, from the very beginning of his work, make love out to be the motive of his life as a researcher.

Rats are the first victims of this strange exploration. For his doctoral research in psychology, Harlow continued the work of his dissertation supervisor, Calvin Stone, who had devoted his scientific career to the food preferences of rats. Harlow undertook the study of choices made by unweaned baby rats: would they prefer cow's milk to other liquids? Would they accept orange juice in the absence of maternal milk? Quinine? Salt-water? To conduct this type of research, the baby rats, of course, needed to be separated from their mother. And from here the story begins. Harlow remarks that the baby rats stop eating when the air is either too cold or too hot. Only a temperature equivalent to that of a mother's body seemed to favor food intake. The food response would thus be encouraged by the fact of being held between the maternal body and the nest. From there it was but a small step to the idea that the babies might perhaps prefer to be with their mother.

A simple step to take, for sure, but for a scientist this step is not taken so easily. Harlow built a cage in which a wire barrier separated mothers and babies. The latter, in desperation, go around in circles in their isolated part of the cage, while the mothers on the other side try to chew their way through the barrier. The strength of this impulse must be put to the test. The experiment is turning into an ordeal. If the mothers are starved, and then the wire barrier is removed with the offer of food, which will the mothers choose? They ignore the food and rush straight for their babies. What is the cause of this strange behavior? Does it consist of a reflex? An instinct? Harlow subjects the rats to these new hypotheses. He removes the mothers' ovaries, blinds them, detaches their olfactory bulbs. Blind, without hormones, and even without smell, the mothers continue to rush straight toward their babies. It might very well consist of love—as if love were not interwoven with odors, images, and hormones. But for Harlow, at any rate, it consisted of a drive with a staggering force: the need for contact.

This is how the story begins, and it's how it begins again a few years later in the early 1950s in the Department of Psychology at the University of Wisconsin–Madison. This time, it's no longer rats but baby rhesus macaques, these great heroes of the laboratory who gave their name to our blood Rh factor. Monkeys are not rats, as we know. It's known even better when a colony needs to be composed for research, as they need to come from India, they cost a lot, and they often arrive in very pitiful states. Their diseases contaminate the others, in an endless cycle. So Harlow decides to create his own colony for himself and, to avoid contagion, isolates the newborns from the moment of their birth. The infant macaques who are raised in this manner are in great health, except for one point: they remain passively seated, rocking back and forth without end, their sad gazes fixed on the ceiling, incessantly sucking their thumbs. Furthermore, when they're put in the presence of a conspecific, they turn their backs to him and let out a terrified scream. Only a single thing seems to draw their attention: the little bits of cloth that cover the floor of their cages. They never stop holding and wrapping themselves with these bits of fabric. The baby macaques had a vital need to touch something soft.

It is therefore this vital need that needs to be studied, dissected, and measured. Harlow will from then on begin to work on building surrogate mothers made of cloth. At the same time, he offers the orphans steel-wired mannequins that dispense milk to them. The baby monkeys avoid the latter, coming close to them only for the time necessary to feed, and attach themselves for hours against the body made of fabric. The need to touch therefore constitutes a primary need; it does not need to be supported by the satisfaction of a need, that of feeding, that was thought to be more fundamental.

The soft mannequin possesses not only a body but also a head with eyes, nose, and mouth. Could love be finally taking bodily form with this face? No, it was still a matter of studying the need to touch. The face is not there to give greater reality to the surrogate; it's there to block the way to another explanation. For this face, on the contrary, has nothing attractive about it. It must not actually attract. The eyes are two red reflectors from a bicycle, the mouth a bit of green plastic, the nose a painted black spot. If the face presented any kind of interest for the baby macaques, one could always retort that it isn't the need to touch that led them to press themselves against the mannequin for hours on end but rather the attractive stimuli of the facial features. Harlow will further prove the effectiveness, and the reassuring function, of the surrogate. How? It suffices to take the surrogate away from them. Panic wins. Another experiment can therefore begin. There are still many more things to take away, or to give, to evaluate the effect of their withdrawal.

Taking, separating, mutilating, removing, depriving. There is something of an infinite repetition in all that I have just related. This experiment of separation does not stop with separating beings from one another but consists in destroying, dismembering, and, above all, removing. As if this is were the only act that could be accomplished. I will not ask you to reread the preceding in order to go back over it, but the true thread that guides this story appears: that of a routine that loses control and becomes mad [*folle*]. Separating mothers and their babies, then separating mothers from themselves, from their own bodies, removing their ovaries, their eyes, their olfactory bulbs—what is known as the model of "breaking down" in science—separating for hygienic reasons, then just for separation itself.

This brings to mind what the psychoanalyst George Devereux noted about the origin of ethnopsychiatry in *From Anxiety to Method in the Behavioral Sciences*.[6] He shows that the indifference of scientists is mainly tied to their inability to tell the difference between a piece of meat and a living being, the difference between those who do not know what is going on and those who do, between a "something" and a "someone." What a valid science of behavior needs, he writes, is not a rat deprived of his cortex but a scientist who has been given back his own. Whether deliberate or not, these two references are not chosen by accident: meat that necessarily comes from an animal and a rat subjected to an experiment of privation are in the contemporary world two main modes of violence in relation to animals. Nonetheless, the primary contrast is not as simple as it first seems. For, if it's a question of what should make the scientist hesitate, then thinking that the piece of meat that he'll spoil with acid comes from an animal that he had to kill, and that he'll have to kill others to provide more pieces of meat capable of reacting with acid, may also lead to hesitation. As for the rat deprived of his cortex and the scientist who needs to be returned his, Devereux clearly translates the process at work: method has taken the place of thought. Devereux's choice of example is also not by accident: experiments of privation or separation—I use these terms interchangeably because they put to work the same process—are exemplary of what they highlight. The method appears in its most grotesque version: a stereotypy that applies the same gesture at all levels, a routine that inhibits any possibility of hesitation.

Rats who ran in mazes, to cite just them, were suspected of not using the faculties of learning—association and memory—that were the object of research; instead, they guided themselves by their own habits (☞ **Laboratory**). They used their bodies, their sensitivities, their skin, their muscles, their whiskers, their sense of smell, and who knows what else. They were therefore deprived, with a systematic spirit that likewise confines to stereotypy. John Watson, the father of behaviorism, removed the rat's eyes, olfactory bulb, and whiskers, which are all essential to this animal's sense of touch, before throwing him into the exploration of the maze. And because the rat didn't want to run the maze or search for the food reward, he starved—just another privation experiment: "he began at once to learn the maze and finally became the usual automaton."[7] But who is the automaton in this story?

This type of routine is not just a fact of the laboratory, for the field was not immunized, and still isn't. While observing a troop of langurs in India, the Japanese primatologist Sugiyama transferred the only male of the group—the male whom he claimed was the dominant male protecting and managing the harem—into another group, which was bisexual. It was a disaster. It was also the discovery of the possibility of infanticide among monkeys (☞ **Necessity**). It's worth noting that this type of practice was common among some primatologists, more specifically among those who seemed to be especially fascinated by hierarchy. I also remember the experiments carried out by Hans Kummer that consisted of transplanting the females of one baboon species, which was organized in a polygynous manner, into another troop, which was organized in a multimale–multifemale mode. How would they adapt?

Some experiments, carried out in particular by the primatologist Ray Carpenter, consisted of systematically removing the dominant male from a troop to observe what results from his disappearance. The social group disintegrates, conflicts become numerous and violent, and the group loses part of its territory to others. Now, it is remarkable that at no moment, in not a single experiment, did the hypothesis of stress caused by the manipulation itself seem to need to be mentioned.

The act of removing the dominant as opposed to another monkey is not without interest. Of course, this corresponds perfectly with the fascination exerted by the hierarchy model in this type of research (☞ **Hierarchies**). But at the same time, according to Donna Haraway, this translates a physiological functionalism of the political body.[8] The social group of monkeys functions like an organism (and the organism functions like a political body): take away the head, and you neutralize what assured the law and order.

But why did the researchers subject their animals to these types of experiments? The answer is rather simple: to see what would happen, like poorly behaved adolescents. Or to put it less simply, because the effects permit an inference of causes. Except that one can never know what in fact "causes," other than by denying the effects of one's own intervention. If Harlow, Carpenter, Sugiyama, Watson, and many more had only considered that, in terms of what "caused" the distress, helplessness, and disorientation of their animals, they ought to have taken into account the effect of the evil intention that ran through the entire dispositive, then they would not have been able to claim anything from their research. Their theories ultimately reflect only one thing: a systematic and blind exercise of irresponsibility.

T

FOR TYING KNOTS

Who invented language and mathematics?

Watana is a preconceptual mathematician. Despite her young age, she has already been the subject of scientific articles and videos, and her work has been shown in an exhibition at Paris's Grande Halle de la Villette. She was born in 1995, in the town of Anvers in Belgium. Rejected by her mother, she was adopted by zoo employees. She then spent some time in Stuttgart, Germany, until her arrival in May 1988 at the menagerie in the Jardin des Plantes in Paris. She belongs to the orangutan species, a species that, until now, has not produced many well-known names in the history of mathematics. No animal has done so, despite a few attempts to at least introduce them into the world of arithmetic.

One can find, in the writings of the eighteenth-century naturalist Charles-George Le Roy, some testimonials by hunters claiming that if they attempted to lure away a magpie to steal her eggs, and if they used a strategy of departing the area by leaving behind one of their own in ambush, it would only work if the number of hunters exceeded four. According to this evidence, the magpies could make out the difference between three and four, but not between four and five.

Over the course of the twentieth century, this ability to count was tested in the laboratories of cognitive scientists. Crows and parrots in particular, though they are far from the only ones, could distinguish between cards on which a certain number of points were drawn. The results, however, were disputed: the animals may not be counting but instead recognizing a kind of gestalt formed by the whole. The ethologist Rémy Chauvin responded to this by claiming that this is how we ourselves proceed most of the time and that mathematical geniuses do not physically have the time to make the calculations proposed to them. It must consist of something else. Is it not said that Japanese owners of koi do not know how many fish live in their pools, for there are so many, but that they can immediately perceive if even one is missing?

Rats have also been subjected, through the model of reinforcement, to the question of knowing whether they can "count." A rat, for example, must demonstrate that he is capable of abstaining from pushing on a lever until a certain number of stimuli signals have been issued.

One may recall the even more famous case of Hans, the Berliner horse who was thought to be capable of solving problems of addition, subtraction, and multiplication and even of extracting square roots. And it's true, there were a number of indications that for a while weighed in favor of this hypothesis, during which the horse, when subjected to an impartial jury in September 1904, showed that he could solve problems to which he gave the answer by tapping numbers with his hoof. The psychologist Oskar Pfungst was charged with solving this case and carried it out efficiently: for a budding scientific psychology, it was difficult to think that a horse could count. Pfungst discovered that, during the tests to which Hans was subjected, the horse was reading involuntary bodily clues from the one asking the questions, indicating the moment when he needed to stop counting. The case was considered closed, although some, like Chauvin, called into question the relevance of these results and considered that the horse could very well have been using gifts of telepathy. Of course, the idea of attributing such humanlike competencies to a horse could seem hardly credible. But some persevered with the idea that even if the horse was probably not counting like we do, his competence was not limited to reading the movements of humans. One finds, in the arguments raised within this controversy, an observation previously made by miners. They have claimed that, while observing horses pulling mining carts, the latter would refuse to leave if the usual eighteen carts were not secured behind them.

Some authors also consider the performances of apes in tests that involve trading as evidence of such competencies in calculation. In these experiments, chimpanzees learned to handle money (or tokens) with which they could pay for supplemental food or services. One can smile, or even feel sorry that he is caught in a system of commerce, or just appreciate the fact that the money endorses the idea that he's working (☞ **Work**). Furthermore, in experimental tests that called for cooperation, it was found that capuchins might refuse to cooperate if they had the feeling that an exchange wasn't fair (☞ **Justice**). Though it isn't arithmetic, they could nevertheless compare orders of magnitude in putting together the premises. Some recent research designed to bring back into question the model of the animal as a "rational economic actor" shows that monkeys—capuchins again—are able to use money in trades and that they "calculate," sometimes rationally, sometimes less so. When the price of a product is reduced, they will opt for the cheaper one. But their choices become "irrational"—more or less according to a certain conception of rationality adopted by the experimenters—when the latter propose transactions in which the capuchins can win or lose part of the purchase. *All else being equal,* they prefer trades that give them the feeling of winning.[1]

The fact that Watana can be considered a preconceptual mathematician comes from another area. Her talent is exercised in geometry. This is the proposition of two researchers who studied her at the menagerie in the Jardin des Plantes: a philosopher, Dominique Lestel, and a philosophical artist, Chris Herzfeld.[2] The story begins when Herzfeld becomes intrigued by the behavior of a young orangutan while taking photos. Watana plays with a piece of string with which she seems to tie knots. Closer observation confirms that this is precisely what she is doing. Her caretaker, Gérard Douceau, supports this hypothesis. Watana has always been attracted to the laces of his shoes, he says, and as soon as the occasion presents itself, she tries to untie them.

Herzfeld then consults the scientific literature to research other cases. There is but a single observed case of apes tying knots. In captivity, conversely, the evidence is more promising. In both rehabilitation sanctuaries and zoos, apes have been seen to undo knots and even on occasion to tie them. But these kinds of observations have little chance of being taken up by scientists, for they are anecdotes. Herzfeld also decides to send out a survey via mail.

In the article that summarizes the results of their research, Lestel and Herzfeld clarify that *today,* that is to say, since the research published by Byrne and Whiten in 1988 on lying among primates, this methodological approach is now considered relevant. A parenthetical remark is necessary here. The obligation whereby they feel required to clarify this point shows the path traveled over the last hundred years—provided that, by "path traveled," you don't hear "improvement" but instead a march according to the expression "two steps forward, three steps back." This clarification alone tells a whole side of the history of animal sciences and the way that rivalries between "modes of knowing" succeeded in disqualifying a considerable part of the resources of what would have constituted its corpus (☞ **Fabricating Science**). Darwin carried out many of his investigations in exactly this way, with techniques set aside and writing queries from all four corners of the world like this: "Have you observed . . . ?" For a good part, the observations that would back up his theories came from amateur naturalists, hunters, dog owners, missionaries, zookeepers, and settlers. The only precaution that he felt compelled to make was to specify that the evidence appeared reliable to him, because it came from someone who was trustworthy. In that era, this guarantee was still sufficient.

But this clarification on the part of the two authors of the article also points to something else that can be seen in every scientific article and that is of interest in the practice of publishing: this clarification translates that every scientist addresses himself or herself to colleagues who are "keeping vigil." This demonstrates one of the modes of reflexivity specific to scientists who must construct their objects and thus, on one hand, take care of their methodologies (as is the case here) and, on the other, ensure the reliability of their interpretations, which are always susceptible to opposition by a competing interpretation: "one could argue that," for example, "simple conditioning could explain this" (☞ **Pretenders**; ☞ **Magpies**). Even before they subject their work to the critiques of colleagues, every scientist must create an imaginary dialogue with them so as to anticipate all of their objections, along the lines of a "distributed reflexivity."[3]

Let's return to Herzfeld's investigation. She received ninety-six responses in reply. Among knotting enthusiasts, one finds talking apes, bonobos, and chimpanzees; but the award goes to orangutans, who accounted for seven of the responses, compared to three bonobos and two chimpanzees. All of them were raised by humans, either in zoos or in laboratories. The overrepresentation of orangutans is not surprising; they weave nests in the wild, and in captivity they seem to like to manipulate and play alone with objects.

Watana is no exception, but she is especially gifted. Lestel and Herzfeld therefore propose that she put her talents to the test. In controlled and filmed conditions, the experimental device consists of offering her some material for knotting and bricolage: rolls of paper, cardboard, pieces of wood, bamboo tubes, string, rope, laces, gardening stakes, and bits of cloth. As soon as the material is provided, Watana begins to knot, using her hands, feet, and mouth. She assembles and knots two ends of string, then makes a series of knots and loops, passes the loops through one another, and inserts bits of cardboard, pieces of wood, or bamboo. She creates a necklace with two strings, puts it around her neck, then throws it up in the air several times. Afterward she collects it and carefully unties it. On other occasions, she uses colored thread, or alternatively, she attaches some rope to the fixed supports of her cage and traces forms with them from one point to another in the space.

In nearly every case, she undoes the work she's carried out. Untying the knots is just as important as the tying itself—and, if it should happen some day that someone has the idea for an archaeology of knots, then orangutans will be out of luck; and isn't this a problem for the archaeology of animals? Like the problem that archaeology faces with the discoveries of women's inventions—collection baskets or baby slings—the artifacts of animals have left very few traces, which hardly honors the fact of their having a role in history, or even having a history itself. It is better to invent weapons.

Lestel and Herzfeld wondered about the motive for Watana's behaviors. It doesn't consist of tools, they claim; tools are usually made for a use, which isn't the case here. The hypothesis of play could be convincing because the activity falls within the area of gratuitous behaviors. And yet, Watana refuses to tie knots with Tübo, her partner in their cage and with whom she usually plays.

The fact that she had the idea to use the accessories of her cage as attachment points, and the way that she experimented with this possibility, guided the researchers' hypothesis. Watana creates forms. And these forms indicate that the pleasure is not just for play but that they're meaningful, that they express an *act of generating forms*. It is, they explain, a sort of challenge to which she responds. She does not take the ropes blindly but rather *thinks* about what she can do with them. "She makes sense of what she does and she takes interest in that."[4] Far from executing her work in a nonchalant or distracted manner, she pays close attention and interrupts herself at times to look at what she has done and what still needs to be done. She therefore puts to work an "exploratory logic" that, according to these researchers, refers to the fact that she explores, in a systematic way, the physical and logical properties in the activity of tying knots.

It is in this sense that Lestel and Herzfeld were able to consider Watana as proof of her entry into the universe of mathematics through the door of preconceptual geometry. Of course, she does not prove theorems; she explores the *practical and geometrical* properties of knots *as such*. She recognizes them as the result of *reversible actions*, she has a *functional representation* of them. And she explores them with her body, therefore enacting what they call an "embodied mathematics."

For the two researchers, "interest for forms in themselves and the research of relevant manipulations to deeper explore their properties are *true beginnings* of mathematical activity."[5]

I have chosen to translate the English term "true beginnings" by *origines* (which the dictionary allows) rather than *fondements,* which is presented as equivalent. The two translations cover a very similar meaning, but if I choose the term *origines,* it's because I will find it again, insistently, a little further on in the last section of the article.[6] The term *origines* returns repeatedly, then, as does a "phylogenesis of reason."[7] Admittedly, the project of a nonhuman epistemology is tempting (and I'm not among those who, as the authors fear, could be disturbed by this), much more tempting than that which obliges us to take an interest in a "story of animal groups." But I'm not sure that this is exactly the type of story that honors them. It consists again, and always, of our own story. Chimpanzee, macaque, and baboon groups have now been observed for several generations—which has led Bruno Latour to say that it's rare to have human groups who have benefited from this kind of attention from their anthropologists—and I believe that this is *their* manner of entering into the story: through the little door, which is sometimes the best way. And one doesn't have to arrive all cleaned up with arms full of promises and gifts. Wanting Watana to be in charge of our origin is not giving the apes a history but instead forces them to follow our own and to be our ancestors.

I would say, in defense of Lestel and Herzfeld, that they no doubt wrote within the rules of the genre and that the fascinating investigation they carried out, with imagination and great audacity, had to make a few concessions to the pressures that weigh on publications and research grants. Clearly the question of origins belongs to the rules of the genre, as it seems to respond to an implicit requirement. Interest yourselves in what you want; but if the origin of our behavior is in question, then this becomes interesting *for us*. We have thus saddled apes with a number of origin stories for which they provide the scenario. Savannah baboons have thus served as proof of our first "coming down from the trees," and chimpanzees of the origin of morality, commerce, and still many more things.

In this regard, the award goes to language, for which an impressive number of behaviors have therefore been studied because they would be at the origin. From start to end, these studies therefore give the singular impression of utter comedy. Even the authors for whom I have the greatest respect do not escape this fascination about the origin of language. Bruce Bagemihl, for example, claims that the symbolic gestures accompanying sexual invitations must have favored language acquisition (☞ **Queer**). It would be one of the origins. Chimpanzees who throw their feces at the heads of researchers (☞ **Delinquents**) are examined in the framework of the same program: the act of throwing, in an intentional manner, either rocks or weapons (though the researchers didn't take the risk of giving any to the chimpanzees) would have favored the development of the neuronal centers responsible for language. Last, but not least, and I'll hold myself to this, the anthropologist Robin Dunbar proposed in 1996 that language appeared as a substitute for social grooming.[8] Social grooming, scientists agree, has the function of maintaining social ties. However, because it can only be exchanged step by step, Dunbar says, grooming can only ensure social cohesion in groups of small sizes. Language takes its place, not as a vehicle of informational content, but as a pragmatic activity of "chattering," as an activity to maintain social bonds: speaking with nothing to say allows one to create or to keep contact with others. Except, of course, and this is the weakness of this theory, that one must imagine chattering as prior to every form of spoken language, and it neglects that, to "chatter," one must already have an entire linguistic imagination.

This slightly manic obsession to research *the* origin of language tends to make me smile. In this respect, I sometimes even have the same amused magnanimity toward people (when we don't have to endure them too often) who eternally come back to the same hypothesis, which becomes the comedic inspiration for comic books, humorous essays (e.g., Jean-Baptiste Botul and his *Métaphysique du mou*), the films of Jacques Tati, and literature.[9] What does the parrot-hero of *Zazie in the Metro* say again, on this and every other occasion? "Talk, talk, that's all you can do."[10]

U
FOR *UMWELT*

Do beasts know ways of being in the world?

The American philosopher William James borrowed a line from Hegel when he wrote "the aim of knowledge is to divest the objective world of its strangeness, and to feel ourselves more at home in it."[1] As an introduction to the theory of the *Umwelt,* two of the words in this sentence could be inverted: the theory of the *Umwelt* has as its aim to divest the objective world of its *familiarity* and to make ourselves feel *less* at home in it. I'll return to this proposition in order to correct it again; but I'll leave it as it is, at least provisionally, because it has the merit of giving a pragmatic take on the theory of the *Umwelt*. It asks for a response to the very practical injunction of Donna Haraway's in that it requires us to learn to encounter animals as if they were strangers, so as to unlearn all of the idiotic assumptions that have been made about them.

The theory of the *Umwelt* was proposed by Jakob von Uexküll, an Estonian naturalist born in 1864.[2] In his work, the term *Umwelt,* which means the milieu or environment, will take on a technical sense to mean the "concrete or lived" milieu of the animal.

The intuition behind this theory is seemingly simple: animals, endowed with sensory organs different from our own, do not perceive the same world. Bees do not have the same perception of colors as we do, and we do not perceive the same scents that captivate butterflies, any more than we are sensitive, as a tick is, to the odor of butyric acid released by the sebaceous follicles of a mammal for whom the tick, hanging from a stem or a branch, lies in wait. Where the theory will take a decidedly original turn is in the way that perception will be defined: it is an activity that fills the world with perceptual objects. For von Uexküll, to perceive is to bestow meaning.[3] Only what is perceived and is important to an organism has meaning, in the same way that something doesn't have meaning if it can't be perceived. There is no neutral object, with no vital qualities, in an animal's world. Everything that exists for a being is a sign that affects or an affect that signifies. Each perceived object—I appeal here to the words that Deleuze offers on this theory—*"effectuates a power to be affected."*⁴ The fact that von Uexküll defines "concrete milieu" and "lived milieu" as equivalent finds its meaning in that the two terms refer to "being captured" [*prises*], captured insofar as the direction proves to be indeterminate; on one hand, the milieu "captures" the animal, and affects it, while on the other hand, the milieu only exists because it is an object in being captured, in the way that the animal confers to the milieu the power to affect.

Why does Tschock, Lorenz's jackdaw, suddenly take no interest in the grasshopper that it coveted just a few seconds before? Because it is now motionless; as such, it is no longer significant, and it no longer exists in the perceptual world of the jackdaw. It exists—it affects—only when it jumps. A motionless grasshopper does not have the signification of "grasshopper." This is also why, according to von Uexküll, so many insects are keen to play dead when faced with a predator. Drawing inspiration from him, it could be said that, in the same way that a spider's web is "for the fly," that it is "flylike," the grasshopper has become "for the jackdaw," that it has integrated within its constitution some of the characteristics of its predator. For von Uexküll, because every event in the perceived world is an event that "signifies," and is only perceived because it signifies, every perception makes the animal a "lender" of meaning, that is to say, a *subject*. More concisely, every meaningful perception implies a *subject*, in the same way that every subject is defined as one who bestows meaning.

If I was interested in the *Umwelt* theory, it's mainly for two reasons: because it seemed like it could make animals look less like idiots and because it had the promise of making scientists more interesting. I expected, following Haraway, that this theory would invite us to consider animals as strangers, as "someones" whose behavior is incomprehensible, and not only ask us to suspend judgment but invite us to be tactful and curious: in what world must *this stranger* live so as to present such ways of being? What affects him or her? What precautions does the situation require?

I have to admit I was disappointed. It's likely in part due to the fact that the theory of the *Umwelt* is mostly fruitful for relatively simple animals, namely, those for whom the number of affects that define them are rather limited, those who are no doubt the most familiar strangers to us. The fact that this theory leads researchers to identify the signs that trigger affects has encouraged them to focus on instinctive, and thus the most predictable, behaviors. With only a few exceptions—which the authors of these exceptions will have to excuse me for not mentioning—it proved to be counterproductive compared to what I expected from it. I was probably expecting too much; the animals appeared to me to be limited to following inescapable routines.

In terms of experimentation, the politeness with regard to strange ways of being soon encountered its limits. In this case, it's probably not the fault of the theory but rather the experimental routines that the theory obviously couldn't defuse.

As proof, I'll note the paradoxical character of this relatively recent research that, in a rather interesting perspective on paying attention to the way that monkeys perceive and are affected by their environments, subjected the monkeys to cognitive tests in locations different from their usual enclosures. It is notable, the authors state, that these captive monkeys (capuchins, as it happens) organize their space really quickly by differentiating social spaces from spaces used for sleeping and eating. The researchers' hypothesis is that each of the spaces could prove "enabling" or "incapacitating" for certain cognitive tasks. The idea is interesting; it entails questioning hasty generalizations. It requires slowing down. The research results on competence among animals could not pretend to teach us something if they were not carefully contextualized, and contextualized by the experiment that the animal does with what is proposed to him. If no generalization is already obvious at the heart of the same enclosure, one can imagine the serious hesitation that the researchers may have when the experimental situation moves to another one, or, for that matter, as concerns the generalization to the same group of animals—not counting the generalization of animals to humans. Let's return to the experiment. The hypothesis that guided the research proved to be important, at least from the point of view of the researchers: confronted with the same task of tool manipulation (of extracting syrup using long sticks from the bottom of tubes closed in a box), the capuchins are much more successful in a space where they usually engage in manipulation activities and less successful when they are in a space where they monitor their environment and devote themselves to social interactions. This seems fairly predictable to me—the hesitation is ultimately not with the location—as it could just as well merit explanations other than that of a facilitation by the meaning captured, in this context, by the tool device (e.g., monkeys would be more distracted in the social space). In the end, the results do not invite one to slow down on generalizations because the very question of the "perceptual context," which is no doubt too general, transforms the capuchins into extras in a scene that hardly concerns them. If this is about their lived world, I fear they are just as confused as the researchers are in terms of what they're after.

This testifies even more to the way that research is organized. It begins with a preliminary step, in a procedure, now classic in this area, and that surprisingly is not itself the subject of questioning: the hierarchical ranks of monkeys are determined by subjecting them to a test with just one bottle of milk, on the grounds that this variable could have a role in subsequent tests. One needs to know who is the "dominant" and who is the "subordinate," for this could have an effect on the results (☞ **Hierarchies**). The monkeys therefore comply in gathering around this bottle, thus entering into competition like the researchers hoped, and, very quickly, the hierarchy of dominance that results from this competition is established. There is a funny combination of lived world, variables, and hierarchy, and there is certainly a blind spot in this story: how do the monkeys experience what they're subjected to with this test of hierarchy? How is it important to them? And how will one know, because they were compelled? If there really is a question that the theory of the *Umwelt* raises, and raises in a relevant way, it is that of knowing what matters to animals. It clearly isn't found here.

But the theory could see happier times if I follow Jocelyne Porcher's proposition when she writes, "The nature of farming is to aim to have two worlds cohabit in the most intelligent way possible."[5] In order that the theory of the *Umwelt* keeps to its promises, it will no doubt need to be displaced from its usual locations. No doubt, too, the fact that the theory's promises can be kept is because it's no stranger to the displacement that wisely distances it from scientists who are bound to the watchwords of doing science and to the imperatives of instinct. For Porcher's proposition invites us to explore situations of domestication or farming as places of intercapture at the heart of which new *Umwelten* are created and over-lapped.[6] They are places that make perceptible the porosity of worlds and the flexibility of those who people them. To make two worlds live together in an intelligent way not only means thinking and connecting with what is required in this cohabitation but just as much in taking an interest in what it invents and metamorphoses into.

Also in this respect, Deleuze was right to insist on the fact that animals are "neither in our world, nor in another, but *with an associated world*."[7] With the cohabitation of beings' *Umwelten* associated to worlds that invent modes of coexistence, one finds oneself dealing with a mobile and variable world, with permeable and shifting boundaries. In terms of this possibility, domestication could be defined as the transformation of what was the proper world of one being by another, or, to put it more accurately, the transformation of a being-with-its-world by another being-with-its-world. Not only are cows no longer wild but there is now attached to them a world of stables, hay, hands that milk, Sundays, human odors, touches, words and cries, fences, paths, and ruts. Attached to them is a world that has modified the list of what affects and constitutes them. The very existence of the lead animal—the one who the farmer counts on to help move the herd—could translate the most articulated point of the coexistence: the lead animal is at the center of a network of trust that ties together her companions and the farmer. She is the bond that ties them. Cows from a herd with a lead animal place their trust in the trust that the lead animal shows with respect to the farmer. If she follows, they all follow. Every sphere of domestication can be explored according to this same mode. Dogs have learned to follow the gaze of beings-with-a-world for whom their gaze matters and affects; they have learned to bark with beings-with-a-world who do not stop talking. And likewise, couldn't one say the same thing about the cat "who walks by himself," as Kipling says,[8] or about pigs who are so sensitive to desires, as Porcher states, or again, about horses who, as beings-with-a-world whose bodies carry and matter [*porte et importe*], attune themselves with beings-with-a-world who form a new body with them?

To think these "beings-with-an-associated-world" who mutually trans-form one another in the adventure of domestication brings us back to James. For, if every being comes with an associated world, the *Umwelten*

of the world of farmers and their animals thus constitute an association
of associated worlds, a composition of beings-with-associated-worlds
who associate together. This is what James called a pluriverse. Worlds
whose coexistence creates, experiences, invents, declines, sometimes as a
composition, sometimes as simple copresence.

This means that my proposal to invert James's statement would only
hold on the condition that every one of the terms be subjected to a sig-
nificantly different understanding, an understanding that actually makes
us discover it anew.[9] *The theory of the* Umwelt *has as its aim to divest the
objective world of its familiarity and to make ourselves feel less at home in it:*
the "to feel ourselves less at home in it" takes on new meaning by taking
into account the task of constructing an "ourselves" and an "at home,"
a *domus* for beings who *compose*. Moreover, if I must seriously consider
that beings are neither in one world nor another but *with* a world, this
means that the term "objective world" itself must also be clarified or,
rather, redefined. For "objective world," within the frames of thought
that we use, might still suggest the existence of an objective world *in itself,*
preexisting as such, unified despite and behind appearances. The world
is not objective in this sense, it is multiple. It isn't subjective either—a
temptation sometimes raised in theory today—for the very idea of this
explosion of subjectivities presupposes that below this exists a world that
carries it and provides stable support. What is at play, then, in this multiple
world is not the fact that a species learns how the other sees the world (as
"subjectivism" would have it) but that it learns to discover which world
is expressed by the other and from which world the other is the point of
view. In light of these specifications, this prompts me to return to James's
original proposition: to know is really to divest of their strangeness these
worlds that form the objective world, so as to learn to inhabit them well
and to build them into a being "at home."

And if the idea of an objective world persists, it's because the world is
continuously in a process of objectification. Lived experience is concrete
because it is lived; everything concrete is lived because it is concrete. The
objective world is in a constant process of multiple objectifications, where
some are well stabilized because they are routinely reactualized (like the
world of the tick whose ways of being are reliable), whereas others are
always in a process of experimentation, transforming affects and ways of

being affected, like those *Umwelten* that are partially connected and whose coexistence metamorphoses the beings who are its expression. This is what James called an achievement, namely, when worlds are well associated, *intelligently associated,* as when farmers and animals are happy, together. Other worlds are still destined to disappear, losing a "whole part of reality" to ontological oblivion. Thus, in a novel that recounts the consequences of the disappearance of orangutans for this world, Éric Chevillard wrote, "The point of view of the orangutan, who didn't count for anything in the invention of the world and who holds in the air the terraqueous globe, with its fleshy fruits, termites, and elephants, this unique point of view to whom we owe the perception of the trills of many songbirds and the first drops of a storm on the leaves, this point of view is no more, you realize. . . . The world suddenly shrinks. . . . It's a whole part of reality that collapses, a complete and articulate conception of phenomena that our philosophy now fails."[10]

V

FOR VERSIONS

Do chimpanzees die like we do?

Every word has many habits and powers; one must always
both conserve and employ them all.

—Francis Ponge, *Pratiques d'écriture,*
ou l'inachèvement perpetual

An article in *National Geographic,* accompanied by a photo, went viral on
the Internet in November 2009 and generated a lot of debate. It related
how some chimpanzees in a Cameroon rehabilitation sanctuary behaved
in an entirely unusual manner when their caregivers presented them
with the body of an especially loved older female who had just died:
they remained silent and motionless for a long time, which is altogether
surprising and improbable for such noisy beings.[1] This reaction was interpreted as a behavior of sadness in the face of death. Do chimpanzees
experience mourning? The debates intensified, of course, and versions of
this story multiplied. "It isn't mourning, only humans know this feeling
as it presumes an *awareness of death.*" The cadaver might upset or frighten
them, but nothing suggests that this fear translates a full awareness that
the chimpanzee will no longer be there. Conversely, some invoke the case
of elephants who remain close to a companion's dead body, leaving some
flowers or grass and giving every appearance of a ritual.[2] Other participants
in this controversy have put forward a fairly common criticism with these
types of questions (☞ **Artists**): the chimpanzees did not learn to mourn *by*
themselves, because it was those responsible for the sanctuary who wanted
to show the cadaver so that, they explained, the chimpanzees would "understand the passing." This behavior is therefore not *real* mourning but a
reaction to those who solicited it from them.[3]

On the contrary, one could respond—as I have done in participating
in this debate—that "soliciting" is exactly the kind of term that should

make us hesitate. The initiative may in fact have provoked the grief, not determined it. The chimpanzees' grief could be "solicited" just like our own grief in the face of death—when we need to learn what it means and it is solicited by those who surround us during such a time—which asks us not to forget the link between *soliciting* and *solicitude*. And, if one extends William James's proposal for a theory of emotions, one could consider that grief in the face of death might receive, as a possible condition of existence (the fact that consolation exists), *solicitude* for it. The sanctuary caregivers are therefore very much "responsible" for the grief of the chimpanzees, in the sense that they took responsibility for guiding the chimpanzees' manner of being affected in such a way that they themselves could *respond*; responsibility is not a cause, it's a way of allowing response.

The question of knowing whether it is actually "real" mourning is not all that interesting, and with this type of question, it isn't clear how to escape it. In the tradition of James's pragmatism, on the other hand, the situation lends itself to asking a more important question: what exactly does this ask us to consider?

The contrast between the two questions—"is it really mourning?" and "what does this mourning ask of us?"—conforms with two types of translation: prose and version.[4] Knowing whether it is "real mourning," whether "this says exactly the same thing," refers to *prose*: a translation where the primary value is accuracy and conformity with an original text. Is it actually "real" mourning in precisely the same way that we ourselves understand it? As I define it, prose—which consists in "having an answer for everything"[5]—makes a choice of being synonymous over being homonymous: the two phrases "human-mourning" and "chimpanzee-mourning" must be exactly the same thing, exactly substitutable one for the other. You can go from one to the other without a hitch, so long as it happens in a direct line, without deviation. Conversely, the translation that works from the question "what does this ask of us?" is in line with the other side of translation, specifically, that of *versions*. The response to this question is not itself a version; it is a vector or even creativity.

The version, as a translation that leads from another language back to its own, assumes, like any translation, some choices. In contrast to prose, however, these choices rest on the principle of a multiplicity of possible meanings, in the range of what is possible by "homonymies": the same word can open up a number of meanings and different senses. Drawing from the way that the philosopher Barbara Cassin suggests that the French language is being shaped by the Greek language, not only can every word and each syntactic operation of the source language receive many meanings in translation but they will be translated in the new language by words and syntactic operations that themselves can have several meanings.[6] The version cultivates these differences and bifurcations in a controlled way—in the same way that we say walking is a controlled way of falling.

Consequently, instead of the prosaic question of whether "human-mourning" overlaps exactly with "chimpanzee-mourning," the version substitutes another, doubled procedure. What are the multiple meanings and homonymies that can possibly provide an account of mourning among humans? And the same question can be asked of chimpanzees: what meanings could exist among them? There isn't therefore a translation from one term to another but a double movement of comparison within each universe of possible meanings as a result of what the other induces. It is in this regard that the anthropologist Eduardo Viveiros de Castro uses the term "equivocation." To translate, he says, is to presume that an equivocation always exists; it is to communicate through differences, differences in one's own language (there are a number of different ways to respond to one and the same term), differences in the language of the other, and differences in the very operation of translation. The two *equivocities* are not superimposable. This is what has led Viveiros de Castro to say that "comparison is in the service of translation" and not the other way around.[7] One does not translate in order to compare, one compares for the sole reason of successfully translating. And one compares differences, equivocations, homonyms. Equivocation is the deployment of versions.

Prose attests to the demand for a unique meaning that has the power, on its own, to impose itself. It itself has the power to impose. Translation in versions, on the other hand, consists in connecting together relations of differences.

While I was writing my book *Thinking Like a Rat,* some scientists, to whom I was presenting the initial research that would lead to the first draft, suggested that I clarify what "thinking" meant before I applied it to animals.[8] This suggestion—and I think this was their intention—was meant to convince me to either use another term for the rat or to restrict the term's meaning so that the two referents, the way that a rat thinks and the way that a human thinks, overlap exactly. Both solutions appeal to prosaic translation. I resisted.

I knew that the problematic term *like* was at the heart of this difficulty, because it assumes an acquired similarity and fixed meaning. In the course of writing, I also contemplated abandoning "like" in order to title the book "thinking with a rat." In the end I didn't do so and, in retrospect, I think I was right in not doing so, precisely because the term *like* provokes discomfort. The term *with* would have been a solution for the very fact that it suggests a coexistence without a hitch. But hitches encourage us to "keep watch." *Thinking with,* of course, entails ethical and epistemological obligations, and these obligations are important to me. But the risk with this term is in not showing the difficulty posed by the fact that the meanings only partially overlap in the best of scenarios, following the work on possible homonyms, work that requires proliferating these homonyms only to partially attune them—work that involves showing the operation of translation itself, the choices made, the shifts in meaning that need to be made to perform comparisons, and the combinations that need to be made to ensure transitions that are always messy. The term *like* therefore had nothing of a cheap equivalence whereby concrete instantiations would be sought. It was, however, to be framed as an operator of bifurcations in our own meanings and a creator of partial [*partielles et partiales*] connections. This ultimately returns to "thinking with a rat," wherein this phrase designates not the occasion for empirically thinking like or with rats but the work that the rats obligate us to do in thinking about "how to think 'like.'"

Prose follows a line, word by word; versions draw a web. "Is this mourning 'really' the same?" is therefore the question of prose. "What does this ask of us?" is not strictly speaking a version, but the question leads to one: what are the multiple meanings available in my language or experience, and what are the meanings that are thought to make sense in the experience of chimpanzees? What are we committed to with the differences between their experiences and those that we know? What work of translation are we obligated to in order to connect them?

"A good translation," de Castro writes, "is one that allows the alien concepts to deform and subvert the translator's conceptual toolbox so that the *intention* of the original language can be expressed within the new one."⁹ Translating isn't explaining, and even less explaining the world of others, but rather testing whether or not what one thinks (or experiences) can apply to what others think (or experience). It is experimenting with "how to think when 'thinking like a rat'?"

Thus, for the question "what are the multiple meanings available in my language or experience?" to translate the *mourning* of chimpanzees, I might discover, for example, that by the end, what is available poses a problem. Chimpanzees test me on my language and my sphere of experience because the definitions of mourning that, to all appearances, "we humans all agree on" do not allow us to pass from our sphere to theirs. It isn't the same mourning. But it is precisely at this moment that the question needs to be opened and not closed. It is the moment for considering the failure of bringing them together as a problem, not of the chimpanzees but of our own versions.

"It doesn't have the same meaning as it does for us" is not a sign of the poverty of meaning among chimpanzees; it indicates our own poverty. Mourning became, in my own cultural sphere, a prose—an orphaned or solitary prose, a term without a homonym, prose too poor to connect with anything, prose that puts our experiences on notice.¹⁰ If we want to therefore take seriously the question "what is asked of us to say that chimpanzees experience a version of mourning?" we must, so as not to exclude chimpanzees from the start, put our own concepts to the test of versions. The work of translation thus becomes a work of creation and fabulation, to resist the assignment of prose.

It all leads to this conclusion: mourning is a bleak future for the dead. It is just as much so for the living. The theories of mourning that are taught by psychologists and relayed in philosophy or secular morality courses are theories that are extremely normative and prescriptive. It is a matter of "work," dealing with it through stages, where people must learn to confront reality, to accept the fact that their dead ones are dead, and to separate their ties with the deceased; it is to accept their nothingness and to replace them with other objects to invest in. It is a *conversion*, to be sure, but of a sectarian kind, a conversion that excludes any other version. A prosaic conversion.

So it must be admitted that "mourning" as we understand it may not be suitable for chimpanzees. "Mourning" assigns the dead to nothingness, and it obliges a choice between "real" relations and "imaginary" relations or "beliefs." It assigns reality to what our cultural tradition defines as real. In order that chimpanzees are recognized as having an awareness of mourning, therefore, they must be well aware that when the dead are dead, they no longer exist *anywhere*, and *forever*, except in the minds of the living. Chimpanzees have no reason to adhere to this hypothesis, not because they are incapable of an awareness of "no longer being," of "nowhere," or "forever"—we ourselves know nothing of these—but because there is no *historical reason* that should have led them to think about it.

When put this way, we can begin to investigate some of the muffled and silenced versions, versions that circulate undercover. These versions are found in places where they're accepted as "imaginary"—this is the condition of their acceptance—such as novels, films, and TV series. And if one persists, one also realizes that many people have entirely different theories on loss and grief and do not at all think that the dead should be considered as no longer having anything to ask of them or of us. But there's nowhere real to cultivate these versions. People therefore learn to give up on them to follow the official and expert modes of usage rather than, needless to say, being seen as bizarre, superstitious, naive, or crazy. Or they do not give them up, but can only do so by asking the question of knowing whether they are bizarre, gullible, or crazy. Or again, they find others who think more or less like them, such as spiritualists or mediums, knowing full well that they appear superstitious, bizarre, or gullible.[11]

So to say that chimpanzees *do not have an awareness of death* can pass from prose to version, from the failure of a prosaic translation (what we can say of ourselves cannot be said of others) to the experimentation with a version (what if we spoke *otherwise* about *us*?). I am not suggesting that chimpanzees will propose a new theory of mourning that will save us; they have already put up enough with what has been asked of them as role models. It would not be translating but appropriating. But they call for us to reactivate our lifeless versions, they oblige us to rethink, they put our prose and versions to the test of translation. If the sanctuary caretakers bore the responsibility for creating the grief that they could console, this does not provide us an origins story—"this is how mourning was born"—but commits us to the possibility of another version, which shows that how one responds to mourning gives it, and solicits from it, its particular form, but that it also constrains it in its forms of response: we are given to mourn in a prosaic way because what allows grief to translate an absence, for us, can only receive different translations in a transgressive, surreptitious fashion. Chimpanzees can change us about our own possibilities of changing.

Translating, according to the mode of versions, thus leads to a multiplication of definitions and what is possible, to make more experiences visible, to cultivate equivocations, in short, to proliferate narratives that constitute us as beings who are sentient, connected with others, and affected. To translate is not to interpret, it is to experiment with equivocations.

On one hand, I mentioned that with prose, one is responsible for the choice of term in view of the truth; with versions, on the other hand, one is responsible for the possible consequences resulting from this choice (☞ **Necessity**; ☞ **Work**). To say, then, of an animal that he is the dominant could require prosaic verification that requires verifying that the animal is indeed the "true" dominant in all situations or that this term in fact covers what the literature confers as its meaning (☞ **Hierarchies**). In terms of versions, the question becomes one of asking what commits us to naming it thus: calling one "dominant" privileges a certain kind of story, solicits a certain kind of attention to some behaviors rather than others, makes imperceptible the relation to other possible versions. The term *dominant* is overly loaded and remains well within the order of versions, but a version with a prosaic tendency; it is always the same story told departing from this term; it constrains the scenario. Thinking of translation in terms of versions imparts to the one who has to choose the relevant term the freedom of abandoning or finding, within the means of his or her language, another term that gives birth to another and more interesting narrative— terms like those of deference, charisma, prestige, or "older," as proposed, respectively, by Thelma Rowell for baboons, Margareth Power for Jane Goodall's chimpanzees, Amotz Zahavi for his babblers, or, finally, Didier Demorcy for wolves that we observed together in a Lorraine regional park.[12] Or again, if it's a matter of males, imposing themselves through force and a reign of terror, that of "socially inexperienced" as opted for by Shirley Strum, showing that these attitudes testify above all to the incapacity of these "dominant" baboons, who fascinated primatologists, to negotiate subtly their positions within the troop.

An interest in versions—as can be seen once these terms are used—is not to make a clean break from the others but to create and make visible the relations that others have silenced or to which they gave another meaning.

Versions: this is, in short, what I am attempting to cultivate in this slightly unusual form of successive texts presented as though detached one from the other and that encourages "taking it from the middle," as similar to abecedaries or dictionaries, books of nursery rhymes or poems.[13] Each story receives and gives, and sometimes not, it's own light in the context in which it is welcomed or called. But each story is clarified differently by others who respond from their own contexts of enunciation and according to the fortuitous ways they are connected. These connections between versions can clarify other ways of considering these stories of practices and animals, of assessing their interest, their repetition, the contradictions they raise, and their creativity—sometimes, I have no doubt, against my own ways of comprehending them. But an achievement in this type of dispositive would actually be to make things less simple, and to stammer in one's reading, as I sometimes do with writing, in laughter or in irritation. In short, to cultivate—as Haraway so accurately does, not without unease or trouble, with anger and humor—contradictory versions that are impossible to harmonize.[14]

W

FOR WORK

Why do we say that cows don't do anything?

Do animals work? The sociologist Jocelyne Porcher, who specializes in animal farming, has made this question the object of her research. She began by asking farmers whether it makes any sense for them to think that their animals collaborate and work with them. The proposition is not an easy one—neither for us, nor for many of the farmers.

The same response pours out: no, it is only people who work, not beasts. Of course, it can be conceded that assistance dogs do, as do horses and oxen that pull loads, and a few others associated with professions: police and rescue dogs, minesweeping rats, messenger pigeons, and various other collaborators. The proposition, however, is acknowledged as barely applicable to farm animals. And yet, throughout the investigations that preceded her research, Porcher heard many stories and anecdotes that led her to think that animals actively collaborate in the work of their farmers, that they do things, that they take initiative in a deliberate way. This led her to consider that work is neither visible nor easily thinkable. It is said without being said, seen without being seen.[1]

If a proposition is not easy, it often means that the answer to the question raised by the proposition changes something. This is precisely what guides this sociologist: if we accept the proposition, it must change something. This question is not posed in her sociological practice "for the sake of knowledge"; it is a pragmatic decision, a question for which the answer has consequences (☞ **Versions**). Rare are the sociologists and anthropologists, she remarks, who have imagined that animals work. The anthropologist Richard Tapper seems to be one of the few to have done so. He considers the evolution of relations between humans and animals as having followed a similar history to those of production between humans themselves. In hunting societies, the relations between humans and animals would be communitarian since the animals are part of the same world as the humans. The first forms of domestication would be akin to forms of slavery. Pastoralism would, according to him, reflect contractual forms of feudalism. With industrial systems, the relation is modeled after modes of production and capitalist relations.[2]

This hypothesis, though welcome, will be rejected by Porcher. It has the merit, to be sure, of opening up the idea that animals work, but at the same time it confines the relations to a singular schema, that of exploitation. Therefore, she writes, "it is impossible to think of a different development."

For what Tapper's reconstruction puts into play is the question of what we inherit. To inherit is not a passive verb, it is a task, a pragmatic act. Heritage is built and is always transformed retroactively. It makes us capable, or not, of something other than simply continuing; it demands that we be capable of responding to, and answering for, that which we inherit. We accomplish a heritage, which means the same thing as saying that we accomplish it through the act of inheriting. In English the term *remember* [se souvenir] can take account of this work, work that is more than just memory: "to remember" and "to re-member" [*recomposer*].[3] To create stories, to make history, is to reconstruct, to fabulate, in a way that opens other possibilities for the past in the present and the future.

What can a narrative—that allows the relations uniting farmers and their animals to be thought—change? To start, it would change the relation to animals and the relation to farmers. "To think the question of work," Porcher writes, "obliges one to consider animals as other than victims or natural and cultural idiots that need to be liberated despite themselves." The allusion is clear. She addresses herself to liberationists, to those who, she says, want "to liberate the world of animals," understood here as "ridding the world of animals." This critique indicates the particular stance that Porcher adopts in her work: that of always thinking about humans and animals, farmers and their beasts, together. To no longer consider animals as victims is to think of a relation as capable of being other than an exploitative one; at the same time, it is to think a relation in which animals, because they are not natural or cultural idiots, actively implicate themselves, give, exchange, receive, and because it is not exploitative, farmers give, receive, exchange, and grow along with their animals.

This is why the question "do animals work and actively collaborate in the work of their farmer?" is important, pragmatically speaking. In the absence of a history, it needs to be addressed today. Addressing this to farmers does not therefore come from a pursuit of knowledge—"what do farmers think about . . . ?"—but from a true experimental practice that Porcher invites them to participate in. If she asks them to think, and she actively asks this of them, it is not to collect information or opinions but to explore propositions with them, to provoke hesitation, to try to experiment, in the most experimental sense of the term: what does this do to think like this? And if we try to think that animals work, then what does "work" mean? How to make visible and speakable what is invisible and rarely thinkable?

I claimed that the proposition of thinking that animals work is not easy. As Porcher learned, it is even more difficult because the only place where she could ask it is precisely the place where the meaning of exploitation alone prevails. In other words, the work of animals is invisible *except in places with a lot of mistreatment of humans and animals*.

In effect, the places where the question of animal work comes to be formulated, there where it is most evident, are the worst places of livestock farming, places of farming as a production, such as industrial farming. Porcher explains this apparent paradox: an industrial farm is the place where animals are the furthest removed and distanced from their own proper world such that "their behaviors acutely appear as inscribed within a relation of work." Humans and animals are engaged in a system of "production at any cost" and of competition that promotes the consideration of an animal as a worker: the animal must "do her job" and is punished when she is seen to sabotage the work (e.g., when a sow crushes her young). Workers in these systems, particularly in intensive pig farming, come to consider their work, Porcher says, as personnel management work; this expression is rarely used, but its implicit suggestion never ceases to be present. They must select the most productive sows from the unproductive ones and verify the capacity of the animals to ensure the desired production. They represent themselves as something like "directors of animal resources," she writes, "as evidence of the diffusion of managerial thinking and the increasing role it places at the heart of animal production sectors" (☞ **Killable**). The animal, therefore, occupies a position akin to an obscure, ultraflexible subproletariat that is exploitable and destructible at will. The distinctive trend of industrialization to move away, when possible, from living labor, which is more costly and always prone to error, is found especially in the use of robotic cleaners that replace humans as well as robotic "boars" that replace pigs to detect when females are in heat.

Conversely, the possibility that animals work in well-treated farms appears more difficult to convey. Admittedly, over the course of the study, and when forced to answer, some would conclude by telling her that perhaps, when "seen from this angle," one could think that animals work. This takes time, it demands a serious play with homonymies, it requires that one confer multiple meanings to anecdotes; it's an experimentation. At the same time, it signals that the problem of animal work takes nothing for granted. Porcher decided, therefore, to focus on the evidence itself and the possibility of making work perceptible. She modified her dispositive. She asked the cows.

Ethology has taught us that some questions only receive an answer if they are posed within concrete conditions, not only such that they allow the questions to be posed but that they make those who pose the questions sensitive to the answer and allow them to grasp the answer when it has the chance to emerge. Together with one of her students, Porcher extensively observed and filmed a cattle herd in a barn and noted every instance where the cows needed to take initiative, respect the rules, collaborate with the farmer, and anticipate the farmer's actions so as to allow him to do his job. Porcher also paid attention to the strategies that the cows invented to maintain a peaceful atmosphere, polite maneuvers, social grooming, and the act of letting a conspecific proceed ahead.

What became apparent was the very reason why the work was invisible: the work did not become noticeable, a contrario, except when the cows resisted or refused to collaborate, precisely because this resistance showed that, when all is functioning well, it is because of an active investment on the part of the cows. For when everything runs well, one doesn't see the work. When the cows go peacefully to the milking robot, when they do not jostle with one another, when they respect the order of turn, when they move away from the robot when its operation is done, when they leave the area to allow the farmer to clean their stall (if they do what is necessary to obey an order), when they do what they need to do so that everything runs smoothly, this is not seen as evidence of their willingness to do what is expected. Everything has the look of something that functions or of a simple *mechanical* obedience (the term means what it sounds like); everything flows mechanically. It is only during conflicts where the order is disrupted, for example, when cows take their turn at the milking robot, or when they do not move out of the way to allow cleaning, or when they go elsewhere than is asked of them, when they avoid their duties, or, quite simply, when they dawdle—in short, when they resist—that one begins to see, or rather to translate differently, these situations where everything functions. Everything functions because they have done everything so that everything functions. Periods without conflict, then, are no longer natural, obvious, or mechanical, for they in fact require from the cows a total activity of pacification where they make compromises, groom one another, and offer polite gestures to one another.

A similar statement, though with some important differences, emerges from the research conducted by the sociologist Jérôme Michalon and his work with animals, mainly dogs and horses, who are enlisted as therapeutic assistants for humans who have physical or psychological difficulties.[4] These animals have a passive, "laissez-faire" attitude, but when things get difficult for them, when they "react," it becomes clear that the collaboration is based on an extraordinary capacity for abstention, an active restraint, a determination to "control" themselves that cannot be seen precisely because they have taken on a look of something "taken for granted."

In Porcher's view, everything that appears to be taken for granted now attests to an entire range of collaborative work—*invisible work*—with the farmer. It was only when paying attention to the many ways that cows resist the farmer, overturn or transgress the rules, dawdle or do the opposite of what is expected of them, that the two researchers were able to clearly see that the cows very clearly understood what they had to do and that they actively invested themselves in the work. In other words, it is through "ill will" that, by contrast, will and good will appear; through recalcitrance that cooperation becomes perceptible; through supposed error or feigned misunderstanding that practical intelligence—a collective intelligence—appears. Work is made invisible when everything functions well, or, to put it differently, when everything functions well, the implication that requires everything to function well is made invisible. Cows cheat, pretend not to understand, refuse to adopt a rhythm that is imposed on them, and test the limits, for reasons that are their own but that, by contrast, highlight the fact that they're participating, intentionally, in work. In this respect, I'm reminded of a remark made by Vicki Hearne, the dog and horse trainer who became a philosopher, who asked why dogs always retrieve a stick but drop it a few feet away from where one awaits. It is one way, she says, of giving to humans a measure of the limit of authority that the dog is willing to concede. It's a quasi-mathematical measure that reminds us that "not everything is taken for granted."

What is it that changes, for the cows, such that this active investment in working together becomes visible? Thinking that farmers and cows share the conditions of work—and, following Donna Haraway, this proposition could be extended to laboratory animals—shifts the way that this question is generally opened and closed.[5] This obliges us to think of beasts and people as connected together in the experiment they are in the process of living and through which they together constitute their identities. This obliges us to consider the way that they mutually respond, how they are responsible in the relationship—here *responsible* does not mean that they must accept the causes but that they must respond to the consequences and that their responses are part of the consequences. If animals do not cooperate, the work is impossible. There are not, therefore, animals who "react"; they react only if one cannot see anything other than a mechanical functioning. In operating this shift, the animal is no longer properly speaking a victim, for, once again, being a victim implies passivity, with all of the consequences attached to this, notably, the fact that a victim hardly arouses any curiosity. It is obvious that Porcher's cows arouse much more curiosity than if she had treated them like victims, because they are more lively, more real, they suggest more questions; they interest us and have the chance of interesting their breeder. A cow who knowingly disobeys is involved in an entirely different kind of relation than a cow who departs from the routine because he is stupid [*bête*] and doesn't know any better; a cow who works is involved much differently than a cow who is the victim of the farmer's authority.

If Porcher's research allows us to maintain that cows collaborate in work, can it still nevertheless be said that they work? Can it be maintained, she asks, that they "have a subjective interest in work"? Does work enhance their sensitivity, their intelligence, their capacity to experience life? This question requires that a difference be made between situations in which only the constraint makes the work visible and those where the animals "do their bit" and make the work invisible. To develop this difference, and to account for what characterizes those farm situations where beasts and humans collaborate, Porcher turns to, and gives an original extension to, the theories of Christophe Dejours.[6]

If human work, as Dejours proposes, can be a vector of pleasure and participate in the construction of our identity, it's because it is a source of recognition. Dejours articulates this recognition in the exercise of two types of judgment: the judgment of the "usefulness" of work, which is made by its beneficiaries, clients, and customers, and the judgment of "beauty," which qualifies work that is well done and comes from peer recognition. A third judgment, Porcher suggests, should be added to these: a judgment of the bond. It is the judgment perceived by the workers as having been given by the animals, a judgment that is brought to bear on the work by the animals themselves. It is not brought to bear on the accomplished work or on the results of production but rather on the means of labor. This judgment is at the very heart of the relation with the farmer; it is a reciprocal judgment through which the farmer and the animals can recognize each other. And it's there that the contrast between the situations can be drawn, between the deadly work and destruction of identities in livestock farming where everyone suffers, and the places where humans and beasts share and accomplish things together. The judgment on the link—or judgment on the conditions of living together—makes the difference between work that alienates and work that creates, even in situations that are radically asymmetric between farmers and their animals.

This story still remains to be told, by re-creating a narrative that makes sense of the present so as to offer a future that is a bit more viable. Not an idyllic story about a bygone golden age but a story that whets the appetite for what is possible, that opens the imagination to the unpredictable and to surprise, a story for which a sequel would be desired. This is what Porcher initiates when she recounts, in the very last lines of her book, a memory from her time when she was herself a goat farmer: "Work was the place of our unexpected meeting, the possibility of our communication, when we belonged to different species who, before the Neolithic, even before Neanderthals, apparently had nothing to say and nothing to do with one another."[7] All is said, and yet nothing is.

X

FOR XENOGRAFTS

Can one live with the heart of a pig?

Gal-KO is a strange being. I only know him vicariously through scientific literature, but I can imagine what he's like. When I think of Gal-KO, he brings to mind the existence of the "Piggies" who the earthlings discover when they arrive on the planet Lusitania in the second volume of Orson Scott Card's Ender's Game series.[1] The Piggies are not human; as their name suggests, they are half-human, half-pig. But they think, laugh, feel sad, love, have fear, and become attached to and care for their own as well as the men and women they neighbor (the xenologists and xenobiologists who are tasked with getting to know them). They speak several languages, depending on whether they're addressing females, a companion, or a human who speaks Portuguese—an entire history is signaled by the use of this language by the earthlings. Neither human nor animal, the Piggies put humans to the test with respect to species and kingdom categories. They speak with trees, and the trees respond, but not with what we would call speech. Trees are their ancestors. In every Piggie, there is the seed of a tree that each will become following the ritual of his body breaking down. A new circle of life thus opens before them.

In a fabulatory way—this is precisely one of the essential resources provided by science fiction—I shall attempt to merge the story of Gal-KO with those that the Piggies develop with humans. They are difficult stories in which the lives of some mean the deaths of others, stories in which humans and Piggies encounter one another and try to be honest with one another but where it isn't always possible, where they live and die by and with one another and attempt to compose and recompose together. At an interplanetary level, they are companion species.

As for Gal-KO, he lives on our planet; he belongs to our present, but it is said that he is our future.[2] He resembles a pig because he is one. But the term "companion species" is applied to him in an entirely new way in the history that unites pigs and humans: he is part human and was invented as such only a short time ago. He has been genetically modified so that our bodies can tolerate the organs that Gal-KO will one day donate to us. He has been reconfigured so that the biological and political boundaries by which our bodies differentiate what is "us" and "not us" will no longer be an obstacle to his gift. The "xenogeneic" antibodies responsible for rejecting a transplanted organ in the cases of species "combinations" have been "inhibited," the researchers write, which gives Gal-KO part of his name: the inhibited gene is that which is coded for galactosyltransferase.

There is one last link for me to draw in this connection with the Piggies: the nucleus of Gal-KO that survives beyond his death becomes part of the life of a human. It is this operation that we call a xenograft.

At the moment, xenografts are only practiced in test trials with chimpanzees. Because the chimpanzees' survival rates have not yet reached beyond a year, this method cannot be used with human beings. It is a strange irony of history that these two beings who, in all of earthly memory, have not had much to do with one another find their fates united in physiology laboratories. Chimpanzees have for a long time held the place now occupied by Gal-KO. Since the 1960s, they appeared to be the favored donor due to their relatedness with human beings. The failures of transplantations encouraged a reconsideration of their usefulness; the argument for their proximity on which its potential rested became at the same time an argument against it. The chimpanzee is really too close, so the baboon will pick up the reins. According to a study conducted by Catherine Rémy, the failure of the transplantation of a baboon heart into the body of a little girl for ten days, in 1984, revived the controversy yet again.[3] The newborn was diagnosed with ventricular hypoplasia and died just a little over one week after the transplantation. Debates ensued almost entirely, Rémy remarks, between journalists and health care professionals, with the only other external participants coming from animal rights activist groups. The latter did not solely focus on the fact that an animal had been sacrificed, for the baby girl had also been a victim of this sacrifice. Other critics also noted this, especially when it was observed that the transplants that had been performed until this point were all done on vulnerable or abnormal people (and had all ended in failure): a disadvantaged, blind and deaf individual living in a caravan, a man of color without income, and someone who was condemned to death. The boundaries of humanity are the sites where the questions of what is and what isn't human are posed the least often. The theme of sacrifice easily takes its place. But this discovery will weigh on the debate; the baboon will no longer be a donor.

The fact that chimpanzees are now the ones who replace humans in preclinical trials speaks to the surprising contradictions that proliferate in these types of studies. As soon as the physiological analogy doubles as a moral analogy, the practice becomes problematic, for it then reconfigures itself along other modes of difference and proximity. Chimpanzees can no longer be donors because they are too similar to us; because they are similar, they can act as a substitute to the receiver. There are multiple ways of being the "same" as the other that do not overlap; there are likely many more ways of being "different."

It is in this complicated game of "same" and "different" that the meaning of Gal-KO's taking the reins can be read. For the proximity of the pig to the human, by comparison to the chimpanzee, proves to be much greater from a physiological perspective. They are as close from the perspective of the dimension of their organs—those of the chimpanzee are too small for human adults—as they are from the point of view of the tolerance to the transplanted organ, thanks to genetic manipulations. On the other hand, moral similarities seem to be, at least ostensibly, entirely excluded.

However, this categorization does not prove to be so simple, for it doesn't exactly follow the contours of the distinction of bodies and beings. On one hand, the scientists who Rémy interviewed speak about their practical work by saying of Gal-KO that he is "humanized." The article by the researchers who genetically reconfigured the pig, moreover, uses the term "humanization." In this context, this designation does not refer to what the pig "is" but is a practical term and technical condition: the quality is not based on "sameness" but on "continuity" and allows the passage of one category of being to another. The term "human" allows this action but is not bound by it.

On the other hand, the field study that Rémy conducts within laboratories adds even more to the contradictions inherent in the limits of categories. What "similar to" means in reference to Gal-KO (almost similar, or human) takes on a different meaning depending on whether it is used by researchers or put into action by animal technicians, or even depending on the different situations. The different meanings coexist, but in a compartmentalized way. This regime of coexistence is visible above all in the contrast of practices, that of animal technicians on one hand and scientists on the other. This is to say, it is visible above all in their actions. Rémy observes, for example, that once an organ is removed from an animal, this animal's body is carefully stitched up *after being euthanized* by the animal caretaker. Through this procedure, the animal keeps his body and becomes neither a carcass bound for the butchery nor waste to be thrown away. He is treated like a "close relative." The body will be eliminated, of course, but only after this treatment that maintains the condition of his body, I would say, like that of a "deceased person" (☞ **Killable**). In this condition, the animal *obliges* and, in particular, obliges actions that slow down and break with usual routines.

Other actions appear that also attest to this need to "pay attention." According to Rémy, animal breeders sometimes address themselves with kindness and compassion to the animal who will be subjected to experimental procedures: "Poor old you," one of them says to a pig headed for the operating room, "we will make something of you!"

According to a division of labor that is sufficiently well marked, the researchers delegate matters pertaining to well-being to the animal keepers, as they expect that they'll "know what is in the mind of the animal." When the animals cause trouble or behave unexpectedly, the "slightest nonconformity" leads to confusion for the researchers, who immediately pass along the management to the animal keepers. This organization of work can also be found in the relations maintained one way or another by the researchers and technicians and in an even more noticeable way when they express themselves through forms of humor. Where the former can at times recognize and at other times openly mock the preoccupations of the latter, with their affects and the care that they manifest, and for the fact that they "can enter the minds of animals," the latter ironically see the researchers' lack of good sense.

The contrast is obviously never as clear in discussions as in interactions, especially because that which I have just described could be tempered by the explicit and frequently claimed desire of the researchers to offer "respectful treatment" of the animal . . . by delegating the burden to the technicians. They also insist on the need for an ethical consideration of animals, a "humane" treatment, that aims to approach that which we give to humans. "The paradox," Rémy writes, "is that this understanding of non-humans results, at least in part, in 'excesses' that have led to certain devices. In other words, it is the appearance of an unprecedented instrumentalization that created the conditions for a definition of the animal as a sensitive and innocent creature." A victim. And its death, a sacrifice.

Can a story be told from this paradox? I don't think so, because I don't trust this decision to think of the animal as a victim in order to oblige us to think, no more than I think that sacrifice is of any assistance here in helping us to do so. The reasons that arise from sacrifice are too loaded and situate the question within an inescapable alternative—which, by the way, will always lead to arguments for a greater good. The tradition of the sacrifice has an interesting history, but it doesn't allow for any further application that would oblige us to think here about Gal-KO and the fact of his presence.

What kind of intelligence should we nurture to live with Gal-KO, now that he is here? And how do we do so? The fact that I appealed to the Piggies and science fiction, which problematize humans and their categories, conveys my difficulty. However, if I refer to them, it's above all because the encounter between the humans and the Piggies presents a series of quite concrete problems for xenologists, problems that demand solutions and that do not allow any of those involved to think of themselves as innocent. How should we address ourselves to them? How can we be honest with them, and honest with ourselves, within situations where our interests are irreconcilable? How should we treat them? None of this excludes conflicts, violence, or betrayals on Lusitania. But this is not obvious. There is no "greater good" that can be evoked, and even less so the good of humanity, of which the Piggies could say that it isn't their problem, because it isn't articulated as one of *their* own problems, and that there is no "greater," except as inscribed within power relations.

Of the stories that are handed to us, to which one could Gal-KO be a sequel? How should we imagine inheriting a story for which we would be responsible?

This story has still to be written. I have neither a script nor any outlines.

If I had to search for some, however, I think that I would try to situate them where *metamorphoses* occur. Gal-KO's fate is tied to the possibility of what our imaginations can cultivate: that of metamorphoses, that is to say, the transformation of beings through the transformation of bodies.

But we need to open up the possibilities of this metamorphosis. On one hand, if I follow Rémy's inquiry, scientists never mention metamorphoses with respect to humans. On the other hand, metamorphoses seem to be limited, for the animals, by the system through which the transformations are thought: that of *hybridization*. If this term holds promises in the perspective of a continuous history that leads toward greater and greater diversity, it doesn't keep any of its promises: hybridization remains a matter of a "combination," thus of the reproduction of certain characteristics of the two "parent" species. Thinking in terms of hybridization forces the rest to give and to impose a binary system—of humanized pigs and the possible reciprocation of pigified humans. Metamorphoses, conversely, retranslate "combinations" into a system of "compositions," a system that remains open to surprise and to the event: "other things" can arise that profoundly modify beings and their relations. Metamorphoses are inscribed within myths as well as inventive biological and political fabulations.

Following the analysis that Donna Haraway gives to the work of the biologists Lynn Margulis and Dorian Sagan, I would begin this fabulation with the biological process of "symbiogenesis." I trust this selection all the more so because they respond to an issue similar to my own: creating other stories that offer a different future to "companion species." For many years now, Margulis and Sagan have studied the development of bacteria. Bacteria, they say, never stop passing genes, in a nonstop traffic of going back and forth, and these exchanges never result in the formation of, according to Haraway, "well-bounded species, giving the taxonomist either an ecstatic moment or a headache."[4] The creative force of symbiosis produced eukaryotic cells from bacteria, and the story of every living thing can be reconstructed by including them within this great play of exchange. Every organism, from mushrooms to plants to animals, has a symbiotic origin.

But this origin is not the final word on the story: "creation of novelty by symbiosis did not end with the evolution of the earliest nucleated cells. Symbiosis still is everywhere."[5] Every form of a more complex life is the continual result of more and more intricate and multidirectional acts of association with, and from, other forms of life. Every organism, they continue, is "the fruit of 'the co-opting of strangers.'"[6]

Co-optation, contagion, infections, incorporations, digestions, reciprocal inductions, becomings-with: the nature of being human, Haraway says, is at its most profound, at its most concrete, at its most biological, an interspecific relation—a process of co-opting strangers. I am carefully keeping in mind the term of origin for *xenograft, xenos.*[7] This term appeared for the first time in *The Iliad* and then again in *The Odyssey.* For the ancient Greeks, it signified a "stranger," not a Barbarian but a stranger to whom one offers hospitality. One whose language is intelligible, who can name and speak of his origin. The common language with Gal-KO is that of genetic code, which also designates its actual origin. Gal-KO is aptly named. Is it a language, a way of naming that prepares us to welcome and to think through metamorphoses? Is it a language that makes us more responsible, and more human, in the sense of being "more committed" in interspecific relations?

It's feared that, for the moment, it isn't. This is feared all the more so because, on one hand, Gal-KO is a product in a series, and series do not ask for a consideration of "how to respond" to this question. On the other hand, if scientists can raise the issue of the modification of what they call the "humanization" of the pig, at no point is this same question—and I mean the same question, that of humanization in the sense of being human *differently*—asked with respect to patients who receive a part of Gal-KO's body.

An inquiry conducted by one of the members of the team of researchers who made Gal-KO, for patients who are recipient candidates, attests to the "out of this world" nature of the research. On the basis of the results, I can easily infer the questions that were posed. The results of our inquiry indicate to us, the researchers say, that some of the patients are prepared to receive the donated organ of Gal-KO but only in an emergency and only insofar as they consider the transplant like "a mechanical piece to be swapped in order to put the whole back into working order," no matter if its origin is human or animal. Others refuse, citing a radical difference between the species: "These patients ask to remain human."

Of course, a third and final category, we are further told, attaches conditions and requests more information. They are not specified, and I'm not even sure that they are not themselves entirely dependent on the way that the recipient candidates were questioned. And I do not see anything that would suggest, in this inquiry, the question of knowing whether this type of research was worth it. The sick are held hostage by the questions addressed to them and, in turn, answer as hostages. The answers to this kind of inquiry lead me to think that it was conducted a bit like an investigation of consumers who are asked to take a position with respect to a product, a product that "might be problematic" but wherein the "problem" is already predefined. This does not exactly stimulate intelligence. The researchers carefully avoided questions that might have made them hesitate, unless these questions never even crossed their minds. Their conclusions are proof that this uncertainty is not part of the protocol: "Far from being a vital organ," they write, "the transplanted human organ comes as a voluntary gift from one human to another and, in this capacity, is carefully invested. Reduced to living animal matter, this certainly simplifies the difficult choices that the transplant patients must resolve, especially the impossibility of being able to thank the one to whom one owes one's life."[8]

I'm not so sure that this is really the dilemma faced by people who survive thanks to an organ donation, which implies the death of another being. Some of the novels and autobiographies that I've read, and that try to provide an account of this experience, seem to tell a slightly more complicated story. For these people, it is not so much a matter of thanking but of taking note of the gift and being worthy of it, of accepting to prolong a life that is no longer one's own, of passing from what has become self and other, from what has become oneself of the other and the other in oneself. Another name for metamorphosis? An accomplishment. The donation is thus inscribed within an inherited story, a story still to accomplish.

Therefore it might be among these stories here that we need to look and think, among those stories that recount how we become human with animals. Among what has become of the *gift* and never stops becoming a gift of our nature. A gift to cultivate and to honor, or, in a more challenging version, a gift that demands: to become who the metamorphosis obliges you to be.

In talking about her cat from her youth, doesn't Jocelyne Porcher write that she developed into a human with her? "Part of my identity . . . comes from the animal world and it's my initial friendship with this cat that gave me access. . . . Animals educate us. They teach us to speak without words, to look at the world with their eyes, to love life."[9] If it were only this, to love life.

Y

FOR YOUTUBE

Are animals the new celebrities?

The very first video uploaded to the YouTube website, on April 23, 2005, showed elephants in their enclosure at the San Diego Zoo. One of the three creators of the website, Jawed Karim, began a guided tour of the zoo with them, which lasted only nineteen seconds in this clip. The video remains with the elephants in this episode. Seen in close-up, Jawed hesitates and finishes by saying that . . . elephants have really long trunks. "That's cool," he adds. Didn't the great philosopher Immanuel Kant, himself writing about pachyderms in his *Natural Geography,* say that "[the elephant] has a short tail covered with long bristly hair, which is used for cleaning tobacco pipes"?[1] He didn't add "that's cool," no doubt because it wasn't fashionable to do so. He would have no doubt thought that it's quite practical, however.

Since then, the success of animals on YouTube hasn't stopped growing, and among Internet amateurs, it evokes the metaphor of a truly viral phenomenon, because the spread of content grows as soon as it touches users. This epidemiological metaphor is obviously not without ambivalence; the theme of contagion may just as easily refer to "infectious disease," to Tardean imitation and endemic habits, or to the uncontrollable proliferation of a resistant and destructive virus.[2] Insofar as this third hypothesis seems to me a little too close to what is stirred up by advocates of conservatism, who are agitated, when it comes to humans, by what they call the narcissistic worship of self-exhibition, and who are probably close to apoplectic when this infatuation is directed toward animals, I won't consider it.

On the other hand, the other two versions (that of contagion that leads to metamorphosis and that of imitation that disseminates new habits) seem to me to actually open up interesting paths to explore. YouTube not only reflects new habits but invents and modifies how these habits are spread. In proceeding this way, I want to make another translation perceptible, inspired by what Bruno Latour proposes about the innovation of the Internet and the creation of avatars.[3] If I take up his analysis and apply it to the proliferation of amateur videos in which humans stage themselves, I'd suggest that these videos are a vector for a previously unseen production of new forms of subjectivity—of new ways of being, of thinking oneself, presenting oneself, and knowing oneself. These video practices can thus be redefined as sites for the invention of a new form of psychology, that is, as practices of knowledge and transformation—in the same way that novels, autobiographies, and diaries have been for their readers. Where else did we learn how to fall in love, if not in novels? How have we been formed by stories about rites of passage? How have we become romantics?

The appearance of these writings in our lives has nevertheless remained, for most of us, relatively implicit. This is no longer the case, Latour says, when the Internet becomes involved. Self-production in videos leaves traces that are not only disseminated in an explicit way, they also provoke comments that themselves subsist and, in their own way, encourage more comments and more productions. New habits are created and their dissemination can be tracked, as if the world of the actors who participate in this extensive network had become an elaborate laboratory for psychological experimentation, a laboratory for the creation and theorization of these habits, these ways of being, of entering into relations, of presenting oneself. Could one guess, from this, that the videos that increasingly bought animals into our collective spaces might be, in view of this experimentation, the site of a new ethological practice? Admittedly, I am not using the term in its usual and narrow sense as a "science of animal behavior" but in a way that returns it to its etymology, *ethos*, and the manners, customs, and habits that tie together beings who share, that is, create together, the same ecological niche. In other words, might it be the case that the proliferation of these videos attests not only to new habits but to the creation of a new interspecific *ethos,* of new relational modalities, that at the same time construct knowledge?

A parallel could be drawn here between these new ways of making animals visible, of addressing them, and the practices of diffusion and knowledge that preceded them, such as with wildlife documentaries. As a sign of the interest they draw, they have more or less multiplied expo-nentially since their invention in the 1960s.

What interests me in constructing this similarity is to evaluate their potential to transform the beings involved and the knowledge that unites them. Animal documentaries have carried out remarkable transformations. They have introduced new habits with respect to animals and sometimes even new *ethea* for researchers. The philosopher of science Gregg Mitman claims that the introduction of new communication technologies will, right from the start, introduce scientists more intimately not only to the world of animal communication but just as much to the industry of mass communication.[4] This dual introduction will have several effects. On one hand, the possibility of mass communication will lead to the creation of previously unseen networks in the practices of the promotion of conserva-tion. It will profoundly modify how scientists present their work. Animals, now the stars of films and TV series, are bestowed with "personalities" and emotions; they become "characters" through whom everyone can share in their experiences. On the other hand, intimate contact with animals, from this moment on, emerges as a research methodology—one that is still largely contested but that in some cases can be seen as legitimate. It is all the more so because the creation of this intimate contact, for the public, proves to be an effective stimulus in raising awareness for endangered animals. This new manner of "doing" and communicating research—once relegated to the margins of popular books—will contribute to a blurring of the boundary distinguishing amateur practices from scientific prac-tices (☞ **Fabricating Science**). For many scientists, including those who played the game, this did not happen easily. Many of them watched with some dismay as their practices were assimilated with those of explorers and adventurers and their animals became seriously anthropomorphized.

In their own right, however, these documentaries had a significant ef-fect on the practices themselves. Not only did they inspire the vocations of some researchers—Jane Goodall and her chimpanzees take the prize in this regard—they also encouraged a certain conception of fieldwork based on what they depicted. If one looks, for example, at the research that marks the scientific history of elephants, one can see that the numbers and statistics that indicate the expertise and authority of the researchers have gradually been replaced by "personal" histories, by films and photos that individualize the animals and that bestow upon them a real status of actors in adventures and experiments. These techniques were at first thought to be better at pleading for their protection; but they became a legitimate method of research. These audiovisual practices, furthermore, had a twofold financial impact for the researchers and the protection of animals: network channels that programmed these films contributed extensively to the financing of research projects, and their broadcasting was much more effective in persuading the public to donate to centers for conservation.

It seems appropriate to consider YouTube videos in the wake of sufficiently similar transformations. Of course, everything is on the Web— but one could say the same thing about documentaries, albeit to a lesser extent. Other modes of knowledge are cultivated; amateurs have taken, or rather, taken over, and this time with unrivaled means of distribution. The animals are part of the cast, even more so than in the documentaries. They are talented beings, remarkable for their heroism, sociality, cognitive and relational intelligence, humor, unpredictability, and inventiveness, and they are now part of everyday life. Of course, these documents do not strictly speaking fall within the domain of evidence; hardly anyone is fooled, as the comments attest; nothing is known about how these images were taken, and one can always suspect deception or the possibility that staging has occurred, with or without the complicity of the animals involved. But nearly all of them speak to the evidence of the image: "someone saw it, and the images are proof."

Some of these videos come from researchers and naturalists, but others do not. It's sometimes difficult to discern them. The boundary between the domains of the amateur and the scientist has become blurred, and some of the animals displayed can in effect take on a double identity. This is the impression that really jumps out when one watches some of the ten most watched videos. On October 21, 2011, for instance, one can find among the latter a parrot named Einstein who rivals that of the psychologist Irene Pepperberg (☞ Laboratory), though he is invested with competencies that are noticeably less academic. Many comments scroll down the page beneath the clips, some of which align, in a more familiar version, with the arguments that swirl around the scientific debates about talking animals: that it's conditioning or training, or, on the contrary, that it's proof of their intelligence and that some animals do know what they're saying, or perhaps it's still due to training, but that each iteration of language proves to be *à propos*.

Another clip shows us some polar bears playing with dogs; their games seem to be pulled straight out of the research of Marc Bekoff, especially as the comments seem to echo the scientific theories that Bekoff proposed (☞ **Justice**). The "Battle at Kruger," meanwhile, shows the heroic rescue of a young buffalo from the claws of lions; clearly recorded by tourists, it nevertheless has the look of a real documentary about the way that buffalos socially organize themselves. And as for a spectacular fight between two giraffes, it is introduced with the following disclaimer: "They dont [*sic*] show you this on the TV."

These videos are everywhere today. They arouse, and are proof of, our interest. They sometimes even translate *some* more or less clearly identified interests. For instance, some are taken up as inspiring examples by religious sites. If you search with the keywords "love and cooperation in living things," you'll witness the rescue of a baby elephant by members of its herd, you'll accompany the exemplary cooperative life of meerkats, and termites will demonstrate how to build a structure together. The commentaries on the sites that you'll visit are sometimes written in a moral register (solidarity is of vital importance) and sometimes with a theological intention (who else but a God could create a world in which such phenomena can be found?). In doing so, these strategic uses of animals reconnect with older versions of natural history—and sometimes even with more contemporary versions, but never in such an explicit way, as when it consists of a moral and political register.

A slightly different comparison can also be considered, this time with hidden cameras and other prank videos that perpetuate amateur practices. A cat playing "hide and seek" with his owner, a dog riding a skateboard, an endangered penguin seeking hospitality from some sailors, some monkeys looting the bags of naive tourists; the proliferation of these videos can be seen as the reinvented legacy of comedy TV shows. The style of some of these YouTube clips seems to be marked by a similar spirit. It's possible that the website links that are sent to me or that I discover in my searches do not represent a rigorous sampling, but it seems like the videos that continue this heritage have gradually become the minority. The animals that are now being filmed are no longer very often the victims of outtakes and other extraordinary mishaps; nor, properly speaking, are they clowns. If they're funny, it's because they are doing surprising things, things that are not expected of them. The unexpected obviously has an anthropomorphic feeling to it; animals do things that are drawn from human action, and the humor, surprise, and amazement are more specifically due to the substitution of the actors involved. This is what makes these clips interesting and arouses enthusiasm: animals teach us about what they are capable of and that we have ignored. Even more, because many of the experiences that are shared on the Web are due to the common work between a human and an animal, from the mutual learning that has developed, from a productive complicity, from a game that has been patiently introduced—a dog and his owner on top of a skateboard, a cat who learns how to surprise his owner who is himself hiding; we learn what *we* are capable of with them. An impressive stock of knowledge could well be established, one that uses other methods and networks than those of science, other manners of questioning and testing animals—knowledge that adds new meaning to "companion species" relations.

Scientific practice, however, is not missing from this production of knowledge.[5] It is often along the margins, but to find it, it suffices to follow the traces left on the Web. Thus, with respect to the elephants who paint in a Thailand sanctuary, and the exploration of possible connections therein, one could begin easily enough with a piece written by Desmond Morris, a scientific specialist on painting among monkeys, who himself visited one of the sanctuaries (☞ **Artists**). The video of the alcoholic monkeys on the island of St. Kitts is accompanied by a commentary that provides rather precise statistics on the distribution of this habit (☞ **Delinquents**). It seems difficult to imagine, however, that researchers could have controlled the consumption of these uncontrollable monkeys by observing their daily rampages along tourist beaches—and yet the video makes it seem as though the observation itself of the monkeys in situ had allowed the observation of their daily consumption. In fact, the numbers did not come from the field; rather, the field is what gave scientists the idea to reproduce the conditions that allowed these observations to be transformed into statistics. Identifying the precise terms used in the commentary is enough: "Monkey," the place of "St. Kitts," and, of course,

"Drunk."[6] On the first page of an online search, three articles appear on this subject, two of which recount the process of the scientists' work: how they got the captive monkeys to drink, to what extent, in what condition, with how many animals, and according to what system of propositions. The statistics that the researchers provide us, therefore, cannot pretend to be about the monkeys at the beach but rather about those who were subjected to the experimental procedure in very precise conditions—and, one has no trouble imagining, they were no doubt quite different. The generalization is made too fast, and the results don't have the necessary robustness—it's a little like trying to establish the use of illicit substances or drugs for the human population by studying this in a prison.

Of course, it will be said that we can look for the conditions that led to the production of these numbers, as I have done. But it should not have to be so complicated. The problem is not solely one of precise translation from one sphere to another. If YouTube can become a site for the production of interesting knowledge, combining amateur practices with scientific contributions, then this *hiatus* between the commentaries on the clips and research as it is conducted ought not to exist. More than just rigor is lost in this hiatus, which is precisely what is of interest in what is called "good scientific popularization" [*bonne vulgarisation*]. What makes the "familiarization" of knowledge interesting and important, insofar as it's worthy of its name, is the explanation of these procedures, the precautions of research, the hesitations of researchers, the living beings who are implicated, the processes that authorize the translation of observations into statistics (and statistics into hypotheses), and the debates into which these hypotheses insert themselves. Not only can these "details"—which are anything but—attest to the fact that scientists can speak in a legitimate way on behalf of those they have questioned but they fit into the narrative scheme that makes science interesting: that of enigmas and inquiries, in short that of thrilling and risky adventures.[7]

Of course, some research will appear for what it is, as not particularly interesting and not especially robust; some scientists thus have everything to fear from this test of visibility and have every interest in keeping the public at a distance. But others can claim a really nice achievement: that of interesting us in, and leading us to love, along with their animals, the scientific adventure that mobilizes them.

Z
FOR ZOOPHILIA
Can horses consent?

In July 2005, the lifeless body of a thirty-five-year-old man, Kenneth Pinyan, was left at the emergency room of the hospital in Enumclaw, a small rural city about fifty kilometers outside of Seattle, Washington. The doctors pronounced him deceased. The friend who dropped him off fled the scene, and an autopsy had to be performed to determine the cause of death. The doctors concluded that there was an acute peritonitis due to the perforation of the colon. The investigation shed light on the circumstances that led to this perforation. Pinyan had been sodomized by a horse. His death was found to be accidental. During a police raid, however, authorities discovered a considerable number of videotapes attesting to the existence of a farm where people paid to have sexual relations with animals. The little community was in a panic.

The prosecutors wanted to arraign Pinyan's friend, a photographer by the name of James Tait, whom investigators had found in the meantime. But they were unable to prosecute Tait for the simple reason that bestiality wasn't illegal in the state of Washington. They had even less recourse when it turned out that Tait was not the owner of the farm where the accident took place; it belonged to a neighbor, and Tait was simply the one who had dropped off Pinyan. Tait received a suspended one-year sentence and a three hundred dollar fine and was forbidden from visiting the farm again, all on the grounds of trespassing on his neighbor's property.

At the same time, in France, Gérard X was accused of *nonviolent* sexual penetration of his pony named "Junior." The criminal charge was not due to the zoophilic act but for animal torture. Gérard X received a one-year suspended sentence, was forced to give up his pony, and had to pay a two thousand euro fine to associations for the protection of animals.

These two cases caused a lot of unrest, fear, and lively debate. In the state of Washington, the response to this case led political bodies into previously unseen complications. They needed to quickly overcome the lack of a law for the criminalization of bestiality. They also needed to ensure that this case didn't solely revolve around the horse.[1] In France, the case of Gérard X rallied together various societies for the protection of animals that became part of plaintiff party. In Europe, however, there has been a clear trend in recent years in returning to older laws. Bestiality, which had been decriminalized in a number of countries, is once again being prosecuted, especially under the guise of new laws that cover "sex abuse."

These two cases have rallied scientists from all sides. In the United States, it is two geographers, Michael Brown and Claire Rasmussen, who have examined the history of Kenneth Pinyan's case. In France, a legal scholar and researcher at France's National Center for Scientific Research, Marcela Iacub, has done so.[2] That the law gets involved is perfectly understandable. But what does geography have to do with this case? From what we may remember from high school or university, geography often boils down to the relatively tedious business of maps, regional distributions, geological layers, mountains, and water flows. I regret being born too early! Over the last several years, geography has taken on a rather surprising look, competing even with a number of other areas of study. During some recent research, I discovered, for instance, that there are such things as ghost geographers, who, of course, study and map out places inhabited by ghosts but who also explore, in all of their complexity, the phenomena that haunt these places. Furthermore, when I asked my friend Alain Kaufman, who is a specialist in the relations between sciences at the University of Lausanne, how geography is different from anthropology today, he responded with a smile: geographers draw maps. I confirmed the accuracy of his response in many of the articles that I consulted. The two geographers who occupied themselves with the case of bestiality were no exception; they included two maps with their account. They show the breakdown, state by state, of the presence or absence of legislation against bestiality, first in 1996 and then in 2005. They do not overlap. Over the course of a decade, repressive laws colonized much of the country. These maps, however, are not merely there to justify the professional identities of the researchers. They are at the heart of what interests them, namely, the political changes surrounding sexuality.

Brown and Rasmussen consider themselves as belonging to a new area of geography: queer geography (☞ **Queer**).[3] Queer geography strives to "diversify the subjects, practices, and politics that have been typically and comfortably discussed in sexuality-and-space studies."[4] But, they say, an "uncomfortable consensus" has brought queer geographers together in the last few years; geography has not been queer enough. To live up to a truly queer project, they write, researchers must "(a) get over our squeamishness and concentrate on fucking . . . , or more broadly the ways that particular sex acts structurate normative power relations; and (b) persistently consider those abject, othered bodies, desires, and places that are comfortably occluded by a focus on gays and lesbians per se."[5] One must, in other words, learn to speak about sexuality in terms of bodies and desires and, above all, resist the temptation to consider sex with animals only in the interpretive categories of human discourse. Thinking of sex *with* animals is a test of the evidences and norms that guide our ways of thinking.

Insofar as the controversy on the subject of zoophilia has not only created a fair bit of distress but has above all borne, and has always borne, the sign of contradictions, unease, and discomfort that sexual relations stir up once they're *explicitly* tied to power relations—this is where our interest should be focused. In this way, Brown and Rasmussen follow the call of Michel Foucault from his writings in the late 1970s and early 1980s: we cannot coherently say yes to sex and no to power, for power has a stranglehold over sex.

Iacub, the French legal scholar, articulates her own critique of the Gérard X case on the basis of this same reference. In the condemnation of Gérard X she in effect sees the confirmation of what Foucault was claiming: the reasons for his condemnation are not those of the past, for this isn't a simple account of puritanism. Iacub cites a passage from Foucault's 1979 interview "Sexual Morality and the Law": "sexuality will . . . be a kind of roaming danger, a sort of omnipresent phantom. . . . Sexuality will become a threat in all social relations. . . . It is on this shadow, this phantom, this fear that the authorities would try to get a grip through an apparently generous and, at least general, legislation."[6] Iacub's argument is based on a contradiction: Article 521.1 of the code, in the name of which the owner of Junior, the pony, was found guilty, is the very same one that permits bullfights, the force-feeding of ducks and geese, and cockfights. According to the *very same* law, then, Gérard X can slaughter and eat his pony if this floats his fancy, but he can't have a good time with him—and as Iacub notes, this was not painful for the animal (which the judge accepted), because the act was considered as having been perpetrated "without violence." At the heart of this verdict, Iacub writes, is the problem of power over sexuality, of power that constitutes sexuality as *the* danger, and, in a related way, of consent: for if the criminal charge is actually that of *nonviolent* penetration, then torture cannot be raised *unless* it is assumed that this penetration was nonconsensual. This means that the question of consent is at the heart of the criminal charge. With respect to the law that is invoked in the conviction of Gérard X, Iacub notes, consent cannot be raised without raising a few contradictions.

In the state of Washington, this legal imbroglio was not so easily resolved, because the contradiction raised by the French legal scholar understandably bogged down the debate. Before making a judgment on subsequent cases that would inevitably arise, legislation was required, and to legislate, a reason was required. The first reason for the implementation of a law was a practical, and seemingly urgent, one: as claimed by the conservative senator Pam Roach, who was put in charge of the case, without a law, the state of Washington risked becoming what could be called a "sex paradise"—or, in the much more loaded words of the senator, the "mecca for bestiality." Thanks to the Internet, and all the rumors surrounding the case, tourism to the farm in this peaceful countryside would take on a strange look and attract perverts from all four corners of the planet. As an argument, however, it was not enough to pass legislation. Another one was adjoined, and it earned great support once it was proposed. Animals cannot consent to a sexual act. They are innocent beings, and they could not want such things. It's a dangerous argument, as Iacub shows so well with respect to the case in France. But before all, in an ironic twist to the story, the argument of consent was contrary to the facts in the Pinyan case. This deserves to be told in greater detail.

Pinyan, in fact, owned a horse who resided on the property of Tait, the friend who accompanied him the night of his death. During that fatal night, Pinyan first made his advances with his own horse. The horse, however, refused to sodomize him, as he wasn't *receptive*, according to the sheriff leading the investigation. Pinyan and his sidekick therefore decided to go over to the neighboring farm where a horse with the eponymous nickname of Big Dick could be found. His reputation was well established. Unbeknownst to the owner, they slipped into the stalls and found the aforementioned Big Dick, who proved to be willing; a little too much so, no doubt, as the rest is history.

Nevertheless, because Senator Roach's concern was not with judging this past case but with establishing a general law for the future, the question of consent seemed to be the best means of argumentation. But now the senator had to face the facts, for the question of consent did not have a place within the legal vocabulary. Animals are listed under the category of property; according to law, property cannot consent, only a holder of property can do so. One cannot be nonconsenting if one is not within the category of beings who can consent. Then there's another problem (and here we come back to Iacub)—the issue is particularly slippery: do animals consent to being held on a leash, to being kept in a zoo, to having drugs tested on them, to being fattened up before being killed and eaten? If, in fact, no one needs to ask their opinion on these matters, it's because they are "property" and thus beings to whom one does not need to legally pose the question of whether they consent.

In light of the difficulties raised by this line of argument, Roach considered another strategy. The scope of the law for the prevention of cruelty needs to be extended to include interspecific relations. Even when extended, however, the law would not be applicable to this case because the physical injury was not to a horse but to a man. Senator Roach considered restricting the question of violence by defining the sexual act itself as abuse. But for this to be abuse, it needs to be shown that the abuser has preyed upon the weakness of the victim. This may cover situations that involve chickens, goats, sheep, or dogs, but not horses. This strategy therefore had to be abandoned, unless one considered every sexual act, a priori, as abuse.

Senator Roach therefore decided to entrust her line of argument to human exceptionalism. This approach was recommended to her by an eminent member of the Discovery Institute, the very conservative think tank and advocate for the neocreationist theory of intelligent design.[7] Bestiality contravenes human dignity; if human exceptionalism is in peril, the law must therefore remind humans of their duty toward their dignity. This argument was not new, but it had to be renewed somehow. Things would have in fact been much simpler if the law against sodomy were still in effect. But because it had been suppressed, it left the door open to bestiality. And because the antisodomy law in this state rested on the fact that sodomy was a "crime against humanity," it "violated human nature."

This unfortunate legislative lacuna did not prevent the law that criminalizes bestiality from finally being promulgated, with one supplementary clause: it would also be forbidden to film this type of practice. Obviously, the legislator gave up on the argument on behalf of human dignity and returned to that on behalf of cruelty. For just four months later, in October 2006, a man was arrested after a complaint made by his wife that he had had sexual relations with their four-year-old female bull terrier. He was convicted for cruelty toward an animal. His wife was successful in having him convicted by presenting to the authorities a video that she was able to record with her cell phone when she came upon them by surprise. To my knowledge, no prosecution was brought against her . . .

The question of consent was raised in the Pinyan case but had to be abandoned. It is at the heart of the French decision, however, and the contradictions that it raises are what caught the attention of Iacub, not only because the contradictions demonstrate the arbitrariness of the verdicts and the law, but also because the emphasis on this notion reveals what is currently taking shape around sexuality. What Foucault anticipated is now happening to us. Sexuality has become the omnipresent danger. The celebrated sexual liberation has become a doctrine, and from now on the state has the power to protect society from a downward spiral. Mutual consent becomes a cornerstone of state control and a weapon for this normalization of sexuality. The very idea of mutual consent, writes Iacub (with the philosopher Patrice Maniglier), will result in a decision as to who is the model sexual victim and who is the compulsory counterpart: the possibility for the state to interfere in individual sexuality to protect the victims. It does so ceaselessly in other areas that have the same arrangement articulated around consent. The act, for example, of characterizing people as manipulable, subjugated, psychologically fragile in the face of influences, amounts to giving the state the authority to protect these people, thus defined, against themselves and against others and to obtain power over all of those circumstances that might be tied to this fragility.

Thus the case of Junior does not stem, as it would have in another era, from a puritanical reaction; it reveals the way that sexuality has become an issue of power and how the state can, through the intervention of judges, protect morality and confine sexuality under the pretext of protecting the victims.

The geographers Brown and Rasmussen similarly take up the question of consent. It was abandoned by the American senator, of course, because she was unable, as it was, to introduce it into law in the scenario of zoophilia. But, these researchers state, the contradictions that are highlighted by zoophilia at the same time emphasize the contradiction inherent to consent itself. Our democracy is based on the participation of those who can "consent." But at the same time, this notion is a formidable weapon of exclusion. Those who cannot consent are excluded from the political sphere. In social contract theories that form the basis for the notion of consent, *even before* determining which parties of the collective have consented to form the community, "the boundaries between those capable of consenting (the citizens) and those outside of the arena of consent (children, foreigners, slaves, and animals, to name a few) must be determined."[8] Social contract theories, Brown and Rasmussen continue, are held out of sight by a *consensual hallucination,* this foundational and scandalous act of democratic communities: a violent and nonconsensual process that results in excluding *beforehand* one party of beings from the community on the paradoxical basis of consent. Thus defined, the boundaries no longer seem to be arbitrarily imposed but are as if self-evident. Obviously, this quite naturally leads to the granting of different ontological statuses to those who are seen as fully autonomous humans and those for whom autonomy, will, consciousness, and capacity "to consent" are lacking. "The animal's inability to consent is justification for condemning the zoophile and yet that same inability to consent is justification for exclusion from other ethical considerations."[9]

Coming from two clearly distinct spheres, and despite their differences, the geographers and legal scholar have followed a common theme. The threads are certainly different, but their crossings translate how they are thinking, on both their behalves, through their practices as researchers and how they're honoring what their practices ask of them: of how their subjects and the problems that they encounter problematize their disciplines, and our ways of thinking more generally, and of how they disrupt the self-evidence and familiarity of the categories, the concepts, and even the tools that permit us to shape them. They are subjects that make one feel unease, discomfort, confusion, or panic, subjects for which there are no simple solutions. These subjects are queer and political because of their power to destabilize. They are all the more so if we think, for example, of the way the philosopher Thierry Hoquet emphasizes their undermining effects when he writes that "zoophilia undoes anthropocentrism" and adds, with a nod toward Plato, "this ancient inclination of cranes who always want to classify themselves separately from other animals."[10] With the legal scholar, zoophilia reexamines categories that seem obvious, such as those of the intimacy of sex, how sexuality contributes to the creation of identity, the issue of consent, and even how being a person—or animal—is defined. With the geographers, the question of boundaries is at the heart of their work. It's not a question of understanding or knowing what zoophilia is but of taking note of what zoophilia does to our knowledge, our tools, our practices, and even our convictions.

In the introduction to her book *La fin des bêtes,* Catherine Rémy links two situations that, to all appearances, look quite different: an ethnomethodological study that Harold Garfinkel conducted with Agnès, a transsexual, and her own investigation about those in charge of killing animals in abattoirs.[11] These studies, Rémy writes, "magnify" the question of boundaries. In becoming a woman, Agnès highlights the constant work of "the controlled exhibition of femininity" and thus that of the institution of sexuality. The killing of animals, as an act, in turn "magnifies" the existence and production of "humanity's boundaries." The individuals ceaselessly carry out a work of categorization that tells us about the practical accomplishments of the boundary between humans and animals.

Zoophilia is a remarkable site for the "magnification" of boundaries, those between acceptable sex and sex that is deemed deviant, those between humans and beasts. Its efficiency is not limited to the latter. Bestiality also highlights the shifting boundary that presides over relations between the countryside and the city. According to many historians, bestiality was more frequently practiced in the countryside and was even accepted, relatively speaking, as sex initiation for adolescents; cities, by contrast, were preserved, hence its decline with urban migration. Today, the two sides of this boundary are reversed, with cities now considered places of all debauchery. Zoophilia also maps the boundaries between nature and culture, not only because it consists of acts that are considered "unnatural" (☞ **Queer**) but because the animals involved—domesticated animals, like horses, cows, goats, sheep, and dogs—continue to keep this boundary under pressure. It follows the lines, ultimately, but the list could stretch from the boundary between those who are granted the ability to consent (informed consent, today) and those who are denied this ability, such as children, animals, the abnormal . . . The responses, sanctions, moral doubts, actions, and laws that zoophilia calls for are part of the process that accomplishes, ratifies, permits, blurs, questions, or undermines the boundaries. This calls to mind what Karim Lapp said to me one day as he was dealing with some questions about urban ecology. "To introduce an animal to the city is to introduce subversion," he said. I don't know if he knew just how right he was.

NOTES

NOTES FOR FOREWORD

1 "W for Work."
2 "F for Fabricating Science," emphasis added.
3 Alfred North Whitehead, *The Concept of Nature* (Amherst, N.Y.: Prometheus Books, 2004), 29, emphasis added.

A FOR ARTISTS

[The subtitle of this chapter, "Bête comme un peintre," plays on multiple meanings of *bête* as "stupid," "idiot," "animal," and "beast," among others. In keeping with other translations of French philosophy, I've usually translated Despret's "bête" as "beast," with the occasional usage of one of the other variants when it best suits the context.—Trans.] The title of this chapter and its opening postscript derive from a lecture that Marcel Duchamp gave at Hofstra University in 1960. Here is a more complete extract: "'Stupid as a painter.' This French saying goes back at least to the time of Murger's *The Bohemians of the Latin Quarter,* around 1880, and was always used as a bit of a joke in conversations. But why must the artist be considered as less intelligent than your average joe? Is it because his technical skills are essentially manual and with no immediate relation to the intellect? Whatever the reason, we generally maintain that the painter has no need of a special education to become a capital 'A' Artist. But these opinions no longer carry much weight today because the relations between the artist and society changed the day when, at the end of the last century, the artist affirmed his freedom." Duchamp, *Duchamp du signe* (Paris: Flammarion, 1994), 236–39. I thank Marcos Mattéos-Diaz for the considerable help he lent in the writing of this chapter.

1 [http://www.dailymail.co.uk/sciencetech/article-1151283/Can-jumbo-elephants-really-paint—Intrigued-stories-naturalist-Desmond-Morris-set-truth.html—Trans.]
2 The videos can be found on sites such as http://www.koreas.com/

and http://www.youtube.com/. I highly recommend that you watch them to understand the enchantment that I'm evoking. You can find a number of sanctuaries in the area of Chiang Mai where elephants have been taken into care. Most of the sanctuaries offer rides on the backs of elephants to tourists, and some of them organize ecotourism accommodations. All of them insist that both humans and elephants must work together to ensure the survival of the animals, whose dietary needs are enormous. During the tourist season, the two elephants who are painting in the videos offer daily shows at the Maetang Elephant Park, roughly fifty kilometers from Chiang Mai.

3 [Despret uses "auteur" here, which carries the sense of the singular author, director, or creative mind behind a work of art. "Author" comes closest to this meaning in this instance.—Trans.]

4 [Despret's use of "agencement" is specific and one wherein she has actively reappropriated the French from its various English translations (e.g., assemblage, arrangement). Her use of "agencement" highlights a preexisting relation that leads toward agency, and with no corresponding English term, the preference is to leave it in the original. See Vinciane Despret, "From Secret Agents to Interagency," *History and Theory* 52, no. 4 (2013): 29–44.—Trans.]

B FOR BEASTS

1 [English in original.—Trans.]

2 [Both here and in "S for Separations," Despret employs this notion of the animal being broken down *(en panne)*. The analogy to a car breaking down (which is heard in the original French), or being broken down, is implicit and intentional: the idea is that animals are actively broken down by experimenters and their scientific practices, much like a mechanical thing can be, and not from a breakdown from within the animals themselves.—Trans.]

3 George John Romanes, *Mental Evolution in Animals* (London: Kegan Paul, Trench, 1883), 225.

4 [Panurge is a character in Rabelais's *Gargantua and Pantagruel* (1532–64), a story in which Panurge throws his sheep into the sea, whereupon the rest of the flock jump into the sea after him. Despret's reference to Panurge in the context of sheep is an allusion to the French expression *mouton de Panurge*: someone or something that imitates another without thought.—Trans.]

5 ["Insight" is in English in the original.—Trans.]

6 See Michael Tomasello, M. Davis-Dasilva, L. Camok, and K. Bard, "Observational Learning of Tool Use by Young Chimpanzees," *Human Evolution* 2, no. 2 (1987): 175–83; Richard Byrne and Anne Russon, "Learning by Imitation: A Hierarchical Approach," *Behavioral and Brain Sciences* 21 (1998): 667–721; and Richard Byrne, "Changing Views on Imitation in Primates," in *Primate Encounters: Models of Science, Gender, and Society*, ed. Shirley Strum and Linda Fedigan, 296–309 (Chicago: University of Chicago Press, 2000).

7 Michael Tomasello's famous article "Do Apes Ape?" can be found in the book *Social Learning in Animals: The Roots of Culture*, ed. Cecilia M. Heyes and Bennett G. Galef Jr., 319–46 (San Diego, Calif.: Academic Press, 1996). ["Do apes ape?" is in English in the original.—Trans.]

8 Byrne and Russon, "Learning by Imitation," 672.

9 Horowitz's response, where she shows that adult humans are worse than chimpanzees, has been published in Alexandra Horowitz, "Do Humans Ape? Or Do Apes Human? Imitation and Intention in Humans *(Homo sapiens)* and Other Animals," *Journal of Comparative Psychology* 117, no. 3 (2003): 325–36.

C FOR CORPOREAL

1 Haraway's analysis of the work of Barbara Smuts can be found in Donna Haraway, *When Species Meet* (Minneapolis: University of Minnesota Press, 2008), 23–24.

2 Ibid., 24.

3 Barbara Smuts, "Encounters with Animal Minds," *Journal of Consciousness Studies* 8, no. 5–7 (2001): 295. Cited in Haraway, *When Species Meet*, 24.

4 [Despret's reference to the etymological sense of *respect* draws from Haraway's use of *respecere*: "Looking back in this way takes us to seeing again, to *respecere*, to the act of respect. To hold in regard, to respond, to look back reciprocally, to notice, to pay attention, to have courteous regard for, to esteem: all of that is tied to polite greeting, to constituting the polis, where and when species meet." Haraway, *When Species Meet*, 19.—Trans.]

5 What Tarde calls interphysiology ought to, according to him, be the foundation of psychology and, more specifically, of an interpsychology. One of Tarde's favorite examples is that of morning glory growing

with a host plant. That interphysiology integrates the repertoire of a host–parasite relation seems to me the best omen, as it prevents us from limiting all of the examples to interpretations of harmonious relations wherein the agreement is self-evident. See Gabriel Tarde, "L'inter-psychologie," *Bulletin de l'Institut général psychologique* 9 (1903): 133–35.

6 Shirley C. Strum, *Almost Human: A Journey into the World of Baboons* (Chicago: University of Chicago Press, 2001).

7 Shirley Strum and Bruno Latour, "Redefining the Social Link: From Baboons to Humans," *Social Science Information* 26, no. 4 (1987): 783–802.

8 Farley Mowat, *Never Cry Wolf* (New York: Bantam Books, 1981).

D FOR DELINQUENTS

1 The videos of the drunk monkeys can be found on YouTube. For the articles concerning alcoholism among the monkeys, one can consult http://www.noldus.com/; for a more detailed account of the conditions of the protocol, see http://www.ncbi.nlm.nih.gov/. With a bit of looking, one can also find a critique of how the results have been presented (☞ **YouTube**).

2 See Jason Hribal, *Fear of the Animal Planet: The Hidden History of Animal Resistance* (Oakland, Calif.: AK Press, 2010).

3 The reference to Robert Musil is taken from his book *Man without Qualities*, 2 vols., trans. Burton Pike (New York: Vintage, 1996), which I discovered in Isabelle Stengers, "Is There a Women's Science?," in *Power and Invention: Situating Science,* ed. Isabelle Stengers and Judith Schlanger (Minneapolis: University of Minnesota Press, 1997), 125.

4 One can find a list of William Hopkins's publications at http://user www.service.emory.edu/~whopkin/. Some of them can be downloaded.

E FOR EXHIBITIONISTS

1 Vicki Hearne, "The Case of the Disobedient Orangutans," in *Animal Happiness* (New York: HarperCollins, 1993), 177.

2 [Despret's use of *l'expérience* has a double connotation here and elsewhere. It can mean both an experiment (e.g., experimenting with how

we form a "we," how we become actors) and experience (e.g., how we experience being part of a "we").—Trans.]

3 [On the etymology of *cum-panis*, see Haraway, *When Species Meet*, 17.—Trans.]

4 ["Agility" in English in the original.—Trans.]

5 Eduardo Viveiros de Castro has written a number of books and articles on perspectivism, and I refer here to "Cosmological Deixis and Amerindian Perspectivism," *Journal of the Royal Anthropological Institute* 4, no. 3 (1998): 469–88. Other works by Viveiros de Castro are referred to in ☞ **Versions**. ["Manioc beer" [*bière de manioc*] is an indigenous type of beer made from cassava (or manioc) root.—Trans.]

6 The testimonials of animal breeders, along with much else, can be found in Despret and Porcher, *Être bête*. [Part of this book has been translated as "The Pragmatics of Expertise," trans. Stephen Muecke, *Angelaki: Journal of the Theoretical Humanities* 20, no. 2 (2015): 91–99.—Trans.]

7 For an exhibitionist of the first order—and that of a laboratory that actively takes into consideration a taste for exhibiting its animal and the spectacular dimension of its device—I refer you to Alex, the parrot of psychologist Irène Pepperberg (☞ **Laboratory**).

F FOR FABRICATING SCIENCE

[This title is one of those cases where I've had to be a bit flexible with my translation to accommodate the abecedary structure. The title of this chapter is "Faire Science," which is often rendered as doing, making, or conducting science, but in consultation with Despret, we've settled on "Fabricating Science" so as to maintain the "F" heading and, more importantly, her use and understanding of the practice of "doing science." Where "faire science" appears below, I've at times used the alternative (and less alphabetically confined) translations.—Trans.]

1 ["Handicap" is in English in the original.—Trans.]

2 ["Pattern" is in English in the original.—Trans.]

3 Eileen Crist, *Images of Animals* (Philadelphia: Temple University Press, 1999). I owe my reading of the contrast between Darwin and Lorenz, as concerns the observations of the peacocks, to her.

4 The fact that animals become predictable is inspired in particular by the analysis that Dominique Lestel gives to the mechanization of

the animal. Lestel considers the loss of the animal's initiative and the dispositives that aim to eradicate any possibility of surprise as related to the same issue; see Lestel, *The Friends of My Friends*, trans. Jeffrey Bussolini (New York: Columbia University Press, forthcoming). I was, however, previously aware of the question of surprise from the work of Émilie Gomart, who uses this notion to further discuss the theory of action, as initiated in the same vein by Bruno Latour, in exploring the way that surprise arises out of the relations between drug users, stakeholders, and political powers. See Émilie Gomart, "Surprised by Methadone," *Body and Society* 10, no. 2–3 (2004): 85–110.

5   The question of the sidelining of amateurs has been expertly analyzed in Marion Thomas's dissertation "Rethinking the History of Ethology: French Animal Behaviour Studies in the Third Republic (1870–1940)." This dissertation was defended at the University of Manchester's Centre for the History of Science, Technology, and Medicine in 2003. I must also mention the important work of Florian Charvolin on the question of the amateur, and especially on an essential dimension that I have not evoked here, namely, passion. See Charvolin, *Des sciences citoyennes? La question de l'amateur dans les sciences naturalistes* (Paris: l'Aube, 2007).

6   For Amotz Zahavi, I recommend the videos of dancing babblers that can be found on the Internet and that include Zahavi calling them, welcoming them, and offering them breadcrumbs (one can search using the terms "babblers" and "Zahavi"). Further to this, I dedicated one of my books to his work; see Despret, *Naissance d'une théorie éthologique: La danse du cratérope écaillé* (Paris: Les Empêcheurs de penser en rond, 1996). The story of the observation of the cheating babbler is but the beginning of an analysis that I wanted to take up again here. [Part of *Naissance* has been translated by Matthew Chrulew as "Models and Methods," *Angelaki: Journal of the Theoretical Humanities* 20, no. 2 (2015): 37–52.—Trans.]

G FOR GENIUS

1   In terms of the bulls connected to the power of the cosmos by their horns, Jocelyne Porcher clarified that she had heard this from the mouth of biodynamic farmers in terms of Rudolph Steiner's "Agricultural Courses." The quote from Michel Ots comes from his book *Plaire aux*

vaches (Paris: Atelier du Gué, 1994), 9. All of my claims with respect to cows are taken in part from the writings of Porcher and in part from a study that we conducted together with respect to farm breeders in 2006, published as Vinciane Despret and Jocelyne Porcher, *Être bête* (Arles, France: Actes Sud, 2007).

2 The fact of feeling "like an anthropologist on Mars," as Grandin puts it, became the title of the book of the same name by Oliver Sacks, in which a chapter is dedicated to Grandin. See Sacks, *An Anthropologist on Mars: Seven Paradoxical Tales* (New York: Picador, 1995).

3 Temple Grandin, with Catherine Johnson, *Animals in Translation: Using the Mysteries of Autism to Decode Animal Behavior* (Orlando, Fla.: Harcourt Books, 2005), 7. I take up in this section part of the analysis I have made of Grandin in my essay "Intelligences des animaux: la réponse depend de la question," *Esprit* 6 (2010): 142–55.

4 Grandin and Johnson, *Animals in Translation,* 8.

5 Ibid., 24.

6 Ibid., 31.

7 C. J. Cherryh, *Foreigner* (New York: DAW Books, 1994).

H FOR HIERARCHIES

1 See http://www.franceloups.fr/.

2 [Cited in Alison Jolly, "The Bad Old Days of Primatology?," in *Primate Encounters: Models of Science, Gender, and Society,* ed. Shirley C. Strum and Linda Marie Fedigan (Chicago: University of Chicago Press, 2000), 78. Jolly is quoting from K. R. L. Hall and Irven DeVore, "Baboon Social Behavior," in *Primate Behavior: Field Studies of Monkeys and Apes,* ed. Irven DeVore, 53–110 (New York: Holt Rinehart Winston, 1965).—Trans.]

3 Alison Jolly, *The Evolution of Primate Behavior* (New York: Macmillan, 1972), 73.

4 On the question of dominance and the state of the controversy in the early 1980s, see Irwin Bernstein's article "Dominance: The Baby and the Bathwater," *Behavioral and Brain Sciences* 4 (1981): 419–57.

5 Donna Haraway, whose writings have inspired me, has worked a lot on the question of hierarchy. See Donna Haraway, "Animal Sociology and a Natural Economy of the Body Politic, Part 1: A Political Physiology of Dominance," in *Women, Gender, and Scholarship (The Sign Reader),* ed.

Elizabeth Abel and Emily Abel, 123–38 (Chicago: University of Chicago Press, 1983). She picks up and develops these questions further in her book *Primate Visions: Gender, Race, and Nature in the World of Modern Science* (New York: Routledge, 1990).

6 Please refer to Thelma Rowell's 1974 article, in which she takes up and clarifies all of the critiques that she addresses on the notion of dominance: "The Concept of Social Dominance," *Behavioral Biology* 11 (1974): 131–54. I have also drawn from the interviews that she gave with me in June 2005, which were conducted for research leading up to the making of the documentary *Non Sheepish Sheep* (dir. Vinciane Despret and Didier Demorcy, 2005), which was prepared for the exhibition *Making Things Public: Atmospheres of Democracy* at the Zentrum für Kunst und Medientechnologie, Karlsruhe, Germany, March 19–August 7, 2005.

7 Shirley Strum's quite negative reactions to these propositions can be found in her book *Almost Human* (Chicago: University of Chicago Press, 1987). Bruno Latour has written a postface for the French 1995 edition.

8 Part of this chapter is inspired by the analyses of Shirley Strum and Linda Fedigan in their introductory chapter "Changing Views of Primate Society: A Situated North American View," in *Primate Encounters: Models of Science, Gender, and Society,* ed. Shirley Strum and Linda Marie Fedigan (Chicago: University of Chicago Press, 2000). Chapter "H for Hierarchies" also takes up some of the points from an article that I've already devoted to this question, "Quand les mâles dominaient: Controverses autour de la hiérarchie chez les primates," *Ethnologie française* 39, no. 1 (2009): 45–55.

9 For the history on the theory of the "hierarchized" wolf, I have benefited from the assistance of two of my students at the University of Brussels, Mara Corveleyn and Nathalie Vandenbussche, who traced the history of this notion. For Schenkel's theories, we consulted Rudolf Schenkel, "Expression Studies on Wolves: Captivity Observations," in *Basel and the Zoological Institute of the University of Basel,* 81–112. This text is not dated and only indicates that it stems from work carried out in 1947. It is worth the read, as one can find all of the usual theoretical affirmations of the theory of dominance. A few pages from a transcribed version can be downloaded on the Internet from http://www.davemech.org/. For Mech's research, I'd recommend this

summarizing article: David Mech, "Whatever Happened to the Term Alpha Wolf?," *International Wolf* 4, no. 18 (2008): 4–8.

I FOR IMPAIRED

1 A summary of research on endogamous fish can be found on a website of amateur fish breeding, http://www.practicalfishkeeping.co.uk/. The authors of this research have written many articles, but in this chapter I have relied on T. Thünken, T. C. M. Bakker, S. A. Baldaud, and H. Kullmann, "Active Inbreeding in a Cichlid Fish and Its Adaptive Significance," *Current Biology* 17 (2007): 225–29. The authors specify that the fish prefer to mate with an apparently "unrelated" fish, which, from the point of view of the controversy, gives reason to those who pretend that there is no attraction to a related fish. If one follows the research, one can see that they refine their theory in a 2011 article published by *Behavioral Ecology*: when tested for their preference for odors (that of a sister's or that of another female's), the large males were more inclined to prefer their sister's. The smaller ones were less "selective," the authors said, because their choices were limited.

2 Stücklin's work on the vole has not yet been published, and I thank him greatly for sending me his writings and allowing me to share them. They stem from a paper he delivered at a conference in 2011, "How to Assemble a Monogamous Rodent: *Ochrogaster* Sociality in Zoology and the Brain Sciences," paper presented at "The Brain, the Person, and the Social," Center of History of Knowledge, ETH Zürich, June 23–25, 2011.

3 [H. T. Gier and B. F. Cooksey Jr., "*Microtus ochrogaster* in the laboratory," *Transactions of the Kansas Academy of Science* 70 (1967): 256–65.—Trans.]

4 [This story comes from Sigmund Freud, "The Interpretation of Dreams," *Standard Edition* 4 (1958): 119–20.—Trans.]

5 [*Tout-terrain* is a term that Despret borrows from the writings of Isabelle Stengers to mean an idea (or concept, theory, dispositive, etc.) that attempts to conquer and handle every field. It may be thought of as either an "all-purpose" or "all-encompassing" theory, but I've translated it as "all-terrain," which is both more literal and closer to the "philosophical equivalent of a military all-terrain Jeep," as Bruno Latour puts it. http://www.bruno-latour.fr/sites/default/files/93-STENGERS-GB.pdf—Trans.]

J FOR JUSTICE

1   The engineer Isabelle Mauz has for several years carried out excit-
    ing sociological work on protected lands. It's to her that I owe the
    framework of my reading, as she's helped me think through conflict
    situations as political situations in which the human agents take seri-
    ously the fact that animals are also political agents. See Mauz, *Cornes
    et crocs* (Paris: Quae, 2005).

2   One can find a detailed history of the end of these trials in an article
    by Éric Baratay, in which he explains the contexts and forms of excom-
    munication and exorcism practices with animals. He shows that these
    do not correspond to a progress toward greater rationality but rather to
    a progressive exclusion of beasts from the community. When animals
    are excluded from the outset from the community, the excommunica-
    tions that, until then, had banished some beasts *in practice* in each case
    no longer had any reason to be. See Baratay, "L'excommunication et
    l'exorcisme des animaux au XVIIe–XVIIIe siècles: Une négociation
    entre bêtes, fidèles et clergé," *Revue d'Histoire Ecclésiastique* 107, no. 1
    (2012): 223–54.

3   The examples of animal trials are taken from Jeffrey St. Clair's preface
    to Jason Hribal's *Fear of the Animal Planet: The Hidden History of Animal
    Resistance* (Oakland, Calif.: AK Press, 2010).

4   The theme of noninnocence and compromise has been developed, in
    particular, in the work of Donna Haraway. See Haraway, *When Species
    Meet*.

5   See Émilie Hache's beautiful book *Ce à quoi nous tenons* (Paris: Les
    Êmpecheurs de penser en rond/La Découverte, 2011), in which she
    extends some of Haraway's work. In this book, Haché considers
    compromises as ways to *compromise oneself* and to come to terms with
    one's principles.

6   Leo P. Crespi. "Quantitative Variation of Incentive and Performance
    in the White Rat," *American Journal of Psychology* 55 (1942): 467–517.

7   Jules Masserman, "Altruistic Behavior in Rhesus Monkeys," *American
    Journal of Psychiatry* 121 (1964): 585. This article is accessible online at
    http://www.madisonmonkeys.com/.

8   Sarah Brosnan and Frans de Waal, "Monkeys Reject Unequal Pay,"
    Nature 425 (2003): 297–99.

9   [See Thelma Rowell, "A Few Peculiar Primates," in *Primate Encoun-
    ters: Models of Science, Gender, and Society,* ed. Shirley C. Strum and

Linda Marie Fedigan, 57–70 (Chicago: University of Chicago Press, 2000).—Trans.]

10 [Despret's use of *un accord* has a double meaning here and elsewhere in this book: that of an *agreement* between the animals that what they're doing is "play" and of an *attunement* between the animals, in the sense of being attuned to the others' gestures.—Trans.]

11 Bekoff's work on animal play has been published in a number of works, but I've relied mainly on "Social Play Behavior: Cooperation, Fairness, Trust, and the Evolution of Morality," *Journal of Consciousness Studies* 8 (2001): 81–90. This article is accessible online at http://www.imprint.co.uk/. The quote by Irwin Berstein concerning the impossibility of scientifically measuring morality is at the heart of the text. I must highlight that the link Bekoff draws between play and justice is more obvious in English thanks to the possibility of using the term *fair,* and that from which it derives, *fairness,* as articulated in his analysis. Barbara Cassin's edited volume *Vocabulaire européen des philosophies* (Paris: Seuil/Robert, 2004) also claims that the term *fair* is untranslatable. The French translation of John Rawls (according to Catherine Audard, in Cassin's collection) has opted for *equity (équité)* to highlight how Rawls's notion of justice is the result of an agreement *(un accorde).* If, by contrast, the French took *fair-play* verbatim to refer to the absence of cheating, to the use of dishonest means, or for the force and respect for the rules of play, it would no longer translate the idea of honesty that the root word *fair* conveys.

12 ["Fair-play" is in English in the original.—Trans.]

13 [See Haraway, *When Species Meet,* 19; see also "C for Corporeal".—Trans.]

14 In addition to Bekoff's propositions, I am greatly inspired by Haraway's *When Species Meet.*

K FOR KILLABLE

1 For the number of animals consumed each year, I consulted http://www.notre-planete.info/ and http://www.petafrance.com/.

2 The numbers on the various human deaths can be found online at http://www.wikipedia.org/, http://www.actualutte.info/, and http://www.sida-info-service.org. See http://www.planetoscope.com/ for real-time statistics on the main causes of mortality.

3 The reference to what "makes a cause" is from Luc Boltanski and

Laurent Thévenot, *De la justification: Les économies de la grandeur* (Paris: Gallimard, 1991).

4 One can read an interesting critique of pragmatic politics with respect to conflict strategies in Erik Marcus's article "Démanteler l'industrie de la viande," *Cahiers antispécistes* 30–31 (2008), http://www.cahiers -antispecistes.org/. In the article, one can find, discussed in a very precise way, the analogy with the abolition of slavery, an analogy that rests on the idea that these conflicts cannot set the ultimate goal as the attainment of perfection.

5 On the treatment of bodies, I refer you to Grégoire Chamayou's *Les corps vils* (Paris: Les Empêcheurs de penser en rond, 2008).

6 On the concept of "sarcophagus," Vialles has profoundly influenced the field of research on the putting to death of animals with her article "La viande ou la bête?," *Terrain* 10 (1988): 86–96. It can be found online at http://www.terrain.revues.org/. See Noëlie Vialles, *Animal to Edible*, trans. J. A. Underwood (Cambridge: Cambridge University Press, 1994).

7 I recently discovered, thanks to Maud Kristen, the existence of a nice empirical test of Vialles's claims. It is a YouTube video ("moedor de porco") in the form of a "hidden camera" scenario and constitutes a real social psychology experiment: in a large open area, a butcher offers some fresh pork sausage to shoppers. He allows them to taste the sausage, then, to insist on its freshness, he proposes that they see how to make them themselves. The shoppers accept to do so, right up to the point where they become aware that the butcher has taken a young living piglet, put it into a box with a crank on its side, closed the lid, and begun to turn the crank. Sausage starts to come out of a hole on the side of the box. The reactions of horror, outrage, and disgust, and the subsequent refusal of the individuals to eat this meat, speak volumes to the mechanisms of forgetting that are necessary in the consumption of meat.

8 The transformation and disassemblage of animals are inspired from the fieldwork of Catherine Rémy, more specifically from the readings she gives to crime novels (e.g., those of Upton Sinclair, Bertolt Brecht, Georges Duhamel) that recount the novelists' own experiences of visiting abattoirs. Cf. Rémy, *Le fin des bêtes: Une ethnographie de la mise à mort des animaux* (Paris: Economica, 2009).

9 [See Hergé's *Tintin in America: The Adventures of Tintin in America*, trans.

Leslie Lonsdale-Cooper and Michael Turner (London: Egmont Children's Books, 1978), 53. "Corned beef" is in English in the original.—Trans.]

10 The citations on data taking the place of thought as well as the proposition to consider dead animals as deceased derive from Porcher's wonderful book *Vivre avec les animaux: Une utopie pour le XXIe siècle* (Paris: La Découverte, 2011).

11 Haraway, *When Species Meet,* 80.

12 The idea that no species is a priori killable and that there is no way for a being to live without another living and dying differentially is developed in Haraway's *When Species Meet.*

13 [The etymological wording derives from Haraway, *When Species Meet,* 163.—Trans.]

14 Judith Butler, *Precarious Life: The Powers of Mourning and Violence* (London: Verso, 2004), 26. Quoted in Cary Wolfe, *Before the Law: Humans and Other Animals in a Biopolitical Frame* (Chicago: University of Chicago Press, 2013), 18.

15 Wolfe, *Before the Law,* 18.

L FOR LABORATORY

1 Vicki Hearne, *Adam's Task: Calling Animals by Name* (New York: Alfred A. Knopf, 1986), 225. [Emphasis added by Despret.—Trans.]

2 Lynch has devoted a number of his writings to laboratory practices, including "Sacrifice and the Transformation of the Animal Body into a Scientific Object: Laboratory Culture and Ritual Practice in the Neurosciences," *Social Studies of Science* 18, no. 2 (1988): 265–89. The guidelines for his analysis can be found in Catherine Rémy's *La fin des bêtes: Une ethnographie de la mise à mort des animaux* (Paris: Economica, 2009).

3 I discovered the story, as recounted by Mowrer, about the mynas that started talking once they were freed from the constraints of the learning process in Donald Griffin's *Animal Minds* (Chicago: University of Chicago Press, 1992).

4 "Speaking, learning, etc., for the wrong reasons" is inspired by one of Isabelle Stengers's propositions through which she demonstrates in *Médecins et sorciers* that one of the concerns of scientific medicine is to distinguish between those patients who are healing for the right, as

opposed to the wrong, reasons. Isabelle Stengers and Tobie Nathan, *Médecins et sorciers* (Paris: Les empêcheurs de penser en rond, 2004). For a revealing version of what this means in practice—and that inspired me to make the link—I refer you to the way that Philippe Pignarre asks us to rethink the placebo effect. Pignarre, "La cause du placebo," May 2007, http://www.pignarre.com/.

5 On the operations of submission and their invisibility, I refer you to Isabelle Stengers's *Sciences et pouvoir* (Paris: La Découverte, 2002).

6 I mention only as an aside that I have dealt with the experiments conducted by Rosenthal and his rats many times, often in a way that is quite critical. See Despret, *Naissance d'une théorie éthologique: La danse du cratérope écaillé* (Paris: Les empêcheurs de penser en rond, 1996), and Despret, *Hans, le cheval qui savait compter* (Paris: Les empêcheurs de penser en rond, 2004).

7 Pepperberg's work with Alex, who is now dead, has appeared in a number of articles and in the book *The Alex Studies* (Cambridge, Mass.: Harvard University Press, 1999).

8 I discovered the scene about the end of the workday with Alex in an online article written by Pepperberg: http://www.randsco.com/.

M FOR MAGPIES

1 Helmut Prior, Ariane Schwarz, and Onur Güntürkün, "Mirror-Induced Behavior in the Magpie *(Pica pica)*: Evidence of Self-Recognition," *PLOS Biology* 6, no. 8 (2008): 1642–50.

2 George Gallup, "Chimpanzees: Self-Recognition," *Science* 167 (1970): 86–87.

3 Joshua Plotnik, Frans de Waal, and Diana Reiss, "Self-Recognition in an Asian Elephant," *Proceedings of the National Academy of Sciences of the United States of America* 103 (2006): 17053–57.

4 Charlotte Thibaut and Thibaut de Meyer, "Les Éléphants asiatiques se reconnaissent-ils? Jouer avec des miroirs," presented in the course "Éthologies et societes," Université Libre de Bruxelles, 2011.

5 [As shown both here and later, an "achievement" *(réussite)* for Despret is distinct from a "success" (also *réussite,* as well as *succès*). Achievements lend greater agency to the animal than does a success, in the sense that one can still achieve something even in a situation that might otherwise be seen as unsuccessful.—Trans.]

6 Prior et al., "Mirror-Induced Behavior in the Magpie," 1648. [Emphasis added by Despret.—Trans.]

7 [The "cognitive Rubicon" is a reference to the crossing of the Rubicon river by Julius Caesar's army, historically noted as a risky transgression of imperial law: having crossed, there was no turning back. The transgression of a "cognitive Rubicon" by the magpies suggests something similar: they passed over a boundary (the "cognitive Rubicon") in self-recognition.—Trans.]

8 On the contamination of competencies, which seem to me remarkable from one area of research to another, I refer you to the catalog of an art exhibit that I helped curate in the Grande Halle de la Villette in Paris. See Vinciane Despret, *Bêtes et Hommes* (Paris: Gallimard, 2007).

9 Prior et al., "Mirror-Induced Behavior in the Magpie," 1642.

10 [See D. J. Povinelli, A. B. Rulf, K. R. Landau, and D. T. Bierschwale, "Self-Recognition in Chimpanzees *(Pan troglodytes)*: Distribution, Ontogeny, and Patterns of Emergence," *Journal of Comparative Psychology* 107 (1993): 347–72.—Trans.]

11 I borrow the term "recalcitrant" from Bruno Latour's commentary on Isabelle Stengers's propositions. This can be found in his preface to Stengers's *Power and Invention* (Minneapolis: University of Minnesota Press, 1998).

12 This reflection on mirrors is an extension of an already published chapter; see Vinciane Despret, "Des intelligences Contagieuses," in *Qui Sont les Animaux?*, ed. Jean Birnbaum, 110–22 (Paris: Gallimard, 2010).

N FOR NECESSITY

1 Sarah Blaffer Hrdy, "Infanticide among Animals: A Review, Classification, and Examination of the Implications for the Reproductive Strategies of Females," *Ethology and Sociobiology* 1, no. 1 (1979): 13–40.

2 In this chapter, I've followed part of Haraway's critical analysis in "A Cyborg Manifesto: Science, Technology, and Socialist-Feminism in the Late Twentieth Century," in *Simians, Cyborgs, and Women: The Reinvention of Women*, 149–81 (New York: Routledge, 1991), as well as what appears in *Primate Visions*. I was able to complete the story of this controversy thanks to the captivating work of Amanda Rees in *The Infanticide Controversy: Primatology and the Art of Field Science* (Chicago: University of Chicago Press, 2009).

3 For a contextualization of research with respect to child abuse, I refer
   the reader to Ian Hacking's *Rewriting the Soul: Multiple Personality and
   the Sciences of Memory* (Princeton, N.J.: Princeton University Press,
   1995).

4 The articles on infanticide among rats include R. E. Brown, "Social
   and Hormonal Factors Influencing Infanticide and Its Suppression in
   Adult Male Long-Evans Rats *(Rattus norvegicus),*" *Journal of Comparative
   Psychology* 100, no. 2 (1986): 155–61; J. A. Mennella and H. Moltz, "Infanticide in Rats: Male Strategy and Female Counter-Strategy," *Physiology and Behavior* 42, no. 1 (1988): 19–28; J. A. Mennella and H. Moltz,
   "Pheromonal Emission by Pregnant Rats Protects against Infanticide
   by Nulliparous Conspecifics," *Physiology and Behavior* 46, no. 4 (1989):
   591–95; L. C. Peters, T. C. Sist, and M. B. Kristal, "Maintenance and
   Decline of the Suppression of Infanticide in Mother Rats," *Physiology
   and Behavior* 50, no. 2 (1991): 451–56.

5 I note in passing that the explanatory leap that researchers take from
   an experimental induction of infanticidal behavior to the idea that it
   is the inductive conditions that are the explanatory cause of behavior
   appears similar to the one that Philippe Pignarre identifies in the confusion between "biology" and what he calls "little biology" with respect
   to the invention of drugs. For example, the fact that a drug appears
   successful in treating depression in no way authorizes the researcher
   to claim to have found *the* cause of depression. Pignarre, *Comment la
   depression est devenue une épidémie* (Paris: La Découverte, 2001).

6 Grandin, *Animals in Translation,* 69.

7 The idea that the mode of knowledge and the production of the
   existence of infanticidal behavior are inextricably linked, and that the
   latter cannot pretend to be "revealed" because it doesn't preexist the
   experiment, is inspired by the work of Isabelle Stengers, *The Invention
   of Modern Science,* trans. Daniel W. Smith (Minneapolis: University of
   Minnesota Press, 2000).

8 Hrdy's first article on the question of infanticide dates from 1979:
   "Infanticide among Animals: A Review, Classification, and Examina-
   tion of the Implications for the Reproductive Strategies of Females,"
   Ethology and Sociobiology 1, no. 1 (1979): 13–40.

9 On the noninnocence of language, more specifically with respect to
   harem, I refer the reader to Haraway's "A Cyborg Manifesto."

O FOR OEUVRES

1 [The word *oeuvre* is sufficiently well known in English to remain un-translated in this chapter heading, but throughout this chapter, I have translated it as "work" (e.g., a work of art); however, it also carries the slightly different connotation of "accomplishment" (e.g., accomplishing a task). In some contexts, then, I have translated oeuvre in the latter sense.—Trans.]

2 [Bruno Latour, *Reassembling the Social: An Introduction to Actor-Network Theory* (Oxford: Oxford University Press, 2005), 58. Latour's notion of *faire faire* is part of a longer expression, "making someone do something" (faire faire quelque chose à quelqu'un). Like the French word *faire, make* has a number of connotations that include creating and producing as well as forcing and causing. *Fait-faire,* as Despret employs it, shares in all of these subtleties.—Trans.]

3 The notion of "theoretical mistreatment" is drawn from Françoise Sironi's work on transgender clinics. Drawing an analogy between what happens to humans and what happens to beasts is always peril-ous; however, insofar as what she describes deals with situations and shrinks who "theorize" those who come to them (and must also aid the individuals looking to undergo their transformation) and discredit them with their suspicious and insulting theories, and thus contribute to their suffering, the analogy can still hold without being insulting. These bewildering theories [*théories abêtissantes*] have concrete effects on animals, whether they are direct (e.g., in laboratories) or indirect, by legitimating thoughtless treatment (e.g., that they are, after all, only beasts, and not geniuses). The adventure of this political clinic that "makes one think" is a highly interesting read. Françoise Sironi, *Psychologie(s) des Transsexuels et des Transgenres* (Paris: Odile Jacob, 2011).

4 Alfred Gell, *Art and Agency: An Anthropological Theory* (Oxford: Oxford University Press, 1989).

5 [The various usages of *fait-faire* in this sentence all play upon the no-tion of the shield's "agency," as in making something happen, making others do something, putting something into action.—Trans.]

6 Gell, *Art and Agency,* 80–81.

7 The 1956 presentation that I refer to, as well as the introductory theories, can be found in the recently republished Étienne Souriau,

Les Différents Modes d'Existence (Paris: Paris Université Presse, 2009). The coauthored preface by Bruno Latour and Isabelle Stengers is important—even essential. Their preface guides the reading, which is at times difficult, and it raises the speculative air that accompanies the adventure of its discovery; it is what first drew me to the issues raised by Souriau.

8 Étienne Souriau, *Le Sens Artistique des Animaux* (Paris: Hachette, 1965). This wonderful little book remains perfectly current, and it is from this book that the different examples of animals are drawn. It has been a profound inspiration in the writing of this chapter.

9 [In the encyclopedic reference *Aesthetic Vocabulary,* edited by Étienne Souriau with Anne Souriau, "instauration" is defined in part as "establishment, foundation (of an institution, a temple). A formal definition, that underlies the notions of duration and stability. However, the Latin sense of *instaurare* and *instauratio* implies the idea of a new beginning, in order to bring to reality what had not been able to previously. Indeed, the idea of instauration implies a dynamic, active experience that finds its completion in an existence. Instauration tends toward a work." Souriau, *Vocabulaire d'Esthétique* (Paris: Presses Universitaire de France, 1990).—Trans.]

P FOR PRETENDERS

1 Thompson's book *The Passions of Animals* (London: Chapman and Hall, 1851) is now fairly rare, though it can be found in a scanned version on Google Books. I have extensively analyzed Thompson's work in a book of mine, whose title was inspired by him: Vinciane Despret, *Quand le loup habitera avec l'agneau* (Paris: Les Empêcheurs de penser en rond, 1999). [The story of the monkey and crows, and also of the orangutan, in this chapter can be found in Thompson, *Passions of Animals,* 355ff.—Trans.]

2 David Premack and Guy Woodruff, "Does the Chimpanzee Have a Theory of Mind?," *Behavioral and Brain Sciences* 4 (1978): 516–26. I thank my student Thibaut de Meyer for bringing this article to my attention.

3 Ibid., 526.

4 The experiment with crows comes from Bernd Heinrich's book *Mind of the Raven* (New York: HarperCollins, 2000).

5 [Thompson's story refers to crows, whereas Heinrich works with

ravens, hence my oscillation between the species in this sentence. "Corvidean" is my translation of *corbésienne.*—Trans.]

6 Thomas Bugnyar and Bernd Heinrich, "Ravens, *Corvux corax,* Differentiate between Knowledgeable and Ignorant Competitors," *Proceedings of the Royal Society of London, Series B* 271, no. 1546 (2004): 1331–36.

7 [*Savoir-vivre* translates as "manners" or "etiquette," but the literal sense alluded to here can be framed as knowing how to live or survive.—Trans.]

Q FOR QUEER

1 Bruce Bagemihl, *Biological Exuberance: Animal Homosexuality and Natural Diversity* (London: Profile Books, 1999), 95. The examples from the Edinburgh Zoo king penguins are drawn from this book.

2 Linda D. Wolfe, "Human Evolution and the Sexual Behavior of Female Primates," in *Understanding Behavior: What Primate Studies Tell Us about Human Behavior,* 121–51 (New York: Oxford University Press, 1991). Cited in Bagemihl, *Biological Exuberance.*

3 This notwithstanding, there is no mention of any reference to Bagemihl's book; it is, however, confirmed by other sources: at the request of the court, the American Psychiatric Association (APA) was called upon during the trial to act as an amici curiae (friend of the court), a form of general council of experts for a given problem. Reference to Bagemihl's book is included in the legal notice as potentially putting into doubt the unnaturalness of homosexuality. I have not had access to this amici curiae, but one of the most virulent homophobes, Luiz Solimeo, refers to it, which leaves me with little doubt of its existence. See http://www.tfp.org/.

4 The arguments that led to the decriminalization of homosexuality following the Lawrence affair can be found at http://www.bulk.resourece.org/.

5 Bagemihl, *Biological Exuberance,* 262. [The idea of the world as "incorrigibly plural" is cited by Bagemihl from Louis MacNeice's poem "Snow."—Trans.]

6 Françoise Sironi, *Psychologie(s) des transsexuels et des transgenres* (Paris: Odile Jacob, 2011), 14–15.

7 Ibid., 229–30.

R FOR REACTION

1 Daniel Q. Estep and Suzanne Hetts, "Interactions, Relationships, and Bonds: The Conceptual Basis for Scientist–Animal Relations," in *The Inevitable Bond: Examining Scientist Animal Interaction*, ed. Diane Balfour and Hank Davis (Cambridge: Cambridge University Press, 1992), 11.

2 Rowell's propositions in terms of habituation are drawn from an interview she conducted when Didier Demorcy and I produced the video *Non Sheepish Sheep*.

3 Information on the work of Michel Meuret can be found in Cyril Agreil and Michel Meuret, "An Improved Method for Quantifying Intake Rate and Ingestive Behaviour of Ruminants in Diverse and Variable Habitats Using Direct Observation," *Small Ruminant Research* 54 (2004): 99–113. In addition to this, Meuret was kind enough to visit me for a few days, at my invitation, during June 2009; this chapter is a result of our conversations.

4 [Despret's distinction here is between *"representatives"* and *"représent-antes,"* respectively. The former refers to the notion that these goats are not representative of goats in general, whereas the latter refers to the fact that some goats can still represent, in a quasi-political sense, themselves to humans.—Trans.]

5 Victoria Horner, Darby Proctor, Kristin E. Bonnie, Andrew Whiten, and Frans B. M. de Waal, "Prestige Affects Cultural Learning in Chimpanzees," *PLOS ONE* 5, no. 5 (2010): e10625.

6 Haraway, *Simians, Cyborgs, and Women*, 201.

7 Ibid., 198.

8 Ibid., 199.

S FOR SEPARATIONS

1 Barbara Smuts, "'Love at Goon Park': The Science of Love," *New York Times*, February 2, 2003, http://www.nytimes.com/2003/02/02/books/review/02SMUTST.html.

2 Deborah Blum, *Love at Goon Park: Harry Harlow and the Science of Affection* (Chichester, U.K.: John Wiley, 2003).

3 Harry Harlow and Stephen Suomi, "Induced Depression in Monkeys," *Behavioral Biology* 12 (1974): 274.

4 Ibid., 275.

5 [Despret is referring to Blum's book *The Monkey Wars* (Oxford: Oxford University Press, 1994), which details the struggles of animal rights activists and the use of monkeys in research activities.—Trans.]

6 George Devereux, *From Anxiety to Method in the Behavioral Sciences* (The Hague: Mouton, 1967).

7 John Watson, "Kinaesthetic and Organic Sensations: Their Role in the Reaction in the White Rat in the Maze," *Psychological Review: Psychological Monographs* 8 (1907): 2–3. Watson is cited in the wonderful little book by the English historian Jonathan Burt, *Rat* (London: Reaktion, 2006), 103.

8 Haraway, *Primate Visions*.

T FOR TYING KNOTS

1 Information on the capuchins who are involved in trading comes from M. Keith Chen, Venkat Lakshminarayana, Laurie R. Santos, "How Basic Are Our Behavioral Biases? Evidences from Capuchin Monkey Trading Behavior," *Journal of Political Economy* 114, no. 3 (2006): 517–37.

2 The photographic work of Chris Herzfeld is in the collection by Pascal Picq, Vinciane Despret, Chris Herzfeld, and Dominique Lestel, *Les Grands Singes: L'Humanité au Fond des Yeux* (Paris: Odile Jacob, 2009). Some of the films documenting Watana were shown at the exhibition *Bêtes et Hommes,* Grande Halle de la Villette, Paris, 2007. A text by Herzfeld is in the exhibition catalog. In terms of the rest, I've consulted Lestel and Herzfeld, "Topological Ape: Knot Tying and Untying and the Origins of Mathematics," in *Images and Reasoning,* ed. Pierre Grialou, Giuseppe Longo, and Mitsuhiro Okado, 147–63 (Tokyo: Keio University Press, 2005).

3 I first paid attention to the question of "distributed reflexivity" in response to Isabelle Stengers and her use of "rival interpretations" in *Invention of Modern Science.* I came back to it when the anthropologist Dan Sperber remarked to me that I had it easy in being critical with respect to artifacts in laboratories because scientists make these types of critiques mutually with one another. I had only to follow them and relay what they're doing. I concede this to him voluntarily (but only partially), for this allows a pragmatic position to be claimed: follow the agents in what they say and do, without constructing "knowledge behind their backs," and without seeing my work as relying on a regime

of denunciation and unmasking ("scientists don't know what they're doing"). But it is precisely because I think that scientists do not know what they're doing in the domain of psychology (human or animal) that I claim to be doing something other than simply relaying the criticism in which they engage between themselves. I would prefer it if I could trust them and not be in the uncomfortable and contradictory position—with respect to my epistemological choices—of denunciation. In this respect, I remain an "amateur" (like Latour, but in a mode that he would certainly find normative), which is to say, someone who likes things and strives to cultivate and better know what she likes, and who can therefore sometimes say, "This has no taste."

4 Lestel and Herzfeld, "Topological Ape," 154.

5 Ibid., 155. [Emphasis added by Despret.—Trans.]

6 ["True beginnings" is in English in the original. I have left the original French for "origines" (origins, beginnings) and "fondements" (foundations, beginnings) due to the context.—Trans.]

7 Ibid., 160.

8 Robin Dunbar, Grooming, Gossip, and the Evolution of Language (Cambridge, Mass.: Harvard University Press, 1996).

9 [Jean-Baptiste Botul, Métaphysique du mou (Paris: Renaud-Bray, 2007). Botul (1896–1947) is in fact a fictional philosopher created by Frédéric Pagès (1950–) and "wrote" a number of works, including The Sexual Life of Immanuel Kant and Soft Metaphysics (or The Metaphysics of the Flabby).—Trans.]

10 [Laverdure is the parrot-protagonist's name in Raymond Queneau, Zazie in the Metro, trans. Barbara Wright (New York: Penguin Classics, 2001). Queneau's novel was also made into a film of the same name by Louis Malle in 1961.—Trans.]

U FOR *UMWELT*

1 William James, The Pluralistic Universe (Lincoln: University of Nebraska Press, 1996), 11. [James's slightly amended quotation of Hegel comes from William Wallace's translation, The Logic of Hegel, Translated from the Encyclopedia of the Philosophical Sciences (Oxford: Oxford University Press, 1873), 335 (§194). I have myself slightly amended James's own retranslation to fit the context better.—Trans.]

2 The theory of the Umwelt can be found in Jakob von Uexküll, A Foray into the Worlds of Animals and Humans, with a Theory of Meaning, trans.

Joseph D. O'Neil (Minneapolis: University of Minnesota Press, 2010). [As is customary, and following Despret, I have kept the term in its original German. The plural of *Umwelt* is *Umwelten*.—Trans.]

3 [The French word *signification*, and its derivatives, often poses challenges for interpreters and translators, as it can mean both "signification" and "significance," with an emphasis on the "signifying" aspect of the sign, as well as "meaning." This is especially so when it comes to von Uexküll, because his writings and thought dwell on how animal lives have meaning, and live in meaningful worlds, and that this is due to the perceptual signs and significations within their worlds. For the purpose of this translation, I have used both "meaning" and "significance" where appropriate to fit the context.—Trans.]

4 Gilles Deleuze and Claire Parnet, *Dialogues II*, rev. ed., trans. Hugh Tomlinson and Barbara Habberjam (New York: Columbia University Press, 2007), 61.

5 Porcher's proposition to consider the world of farming as a world where *Umwelten* cohabit can be found in *Vivre avec les animaux: Une Utopie pour le XXIe siècle* (Paris: La Découverte, 2011).

6 [Despret's use of *entre-capture* (intercapture) draws from Gilles Deleuze and Félix Guattari's notion to connote a "double capture" of beings in a joint becoming, as Alberto Toscano puts it in "Capture," in *The Deleuze Dictionary*, rev. ed., ed. Adrian Parr (Edinburgh: Edinburgh University Press, 2010), 45. See also Deleuze and Guattari, *A Thousand Plateaus*, trans. Brian Massumi (Minneapolis: University of Minnesota Press, 1987), 10.—Trans.]

7 Deleuze and Parnet, *Dialogues II*, 60–61. [Emphasis added by Despret to "an associated world."—Trans.]

8 Rudyard Kipling, *Just So Stories, for Little Children* (Oxford: Oxford University Press, 2009).

9 The idea that the objectivity of the world is not made through an attunement of points of view but due to a multiplicity of worlds expressed by beings (and not just interpreted) was inspired by Eduardo Viveiros de Castro's writings on Amerindian perspectivism. I'm not attempting to make an Amerindian of von Uexküll; I simply find a pragmatic solution in perspectivism, where we refuse mononaturalism so as to consider situations that I believe outreach the capacities of mononaturalism, or the superficial solution of diverse subjectivities bursting around a unified world—which is one way of not taking the existence of these worlds seriously. The idea of a world in a process

of objectification reflects the way that my becoming a philosopher was sustained with James and with the writings of Bruno Latour, especially his great and wonderful *An Inquiry into Modes of Existence,* trans. Catherine Porter (Harvard, Mass.: Harvard University Press, 2013).

10 Éric Chevillard, *Sans l'orang-outan* (Paris: Minuit, 2007), 18.

V FOR VERSIONS

1   For the photo of the chimpanzees at the Sanaga-Yong Chimpanzee Rescue Centre in Cameroon (*National Geographic,* November 2009) and a discussion of culture, one may refer to the site http://www .cognitionandculture.net/.

2   See also the wonderful chapter on grieving birds in Thom van Dooren's *Flight Ways* (New York: Columbia University Press, 2014), which appeared after the publication of the French edition of this book; it raises the question of other forms of mourning in birds.

3   In terms of "mourning" among chimpanzees, I remember the beginning of a conversation I had with Élisabeth de Fontenay after I had used the term "deceased" [*décédé*]. She told me to use "dead" instead. "They are dead." I know that de Fontenay is more than attentive to the words we use, and that the question of translation is essential to her work. Cf. Élisabeth de Fontenay and Marie-Claire Pasquier, *Traduire le parler des bêtes* (Paris: L'Herne, 2008). It was not a matter of her denying or refusing experiences to animals because they are our own (her attention to everything written about the "silence of beasts," which also happens to be the title of one of here books, is proof of this). In her book on translation, she has written a wonderful chapter—on a text by Marguerite Duras—on the grief of Koko the gorilla who claimed (in the sign language taught to her) to sometimes be sad "without knowing why." I can only speculate about de Fontenay's reticence to use the term "deceased," as I had done, in favor of "dead." Our discussion, when interrupted on this point here, didn't allow us to address our arguments. Her position didn't seem to be determined by the fact that it was about animals; rather, it translated a relation to the dead that prescribes us, in our secular tradition, to maintain a "coherent" [*lucide*] relation with death. It is not death *itself* that interests me but rather the relations possible with those who have passed away. This chapter can begin a follow-up to our discussion . . .

4   The contrast I draw between prose and version is not between prose
    "in itself" and version "in itself." It is a contrast between two man-
    ners in which "common" experience is tested. For an average student,
    "prose" represents the difficult work of saying, in exactly the same
    way, the exact same thing, and this passage is carried out without any
    means of freedom or sensitivity that one can draw from one's own
    language. Barbara Cassin, who drew my attention to this, has shown
    that one can do what I'm calling "versions" when translating one's
    own language into the language of another, that is, when one does
    "prose." At any rate, being able to choose and cultivate homonymies
    in the language of another shows that the procedure comes from what
    I'm calling "versions." Laurence Bouquiaux, who was kind enough to
    assist me with my early drafts, made the same remark. She reminded
    me of how Leibniz nicely proposed that one of the ways of "resolving"
    a controversy could be to go through the requirement of translating
    the problem into the language, or the terms, of another. He did it so
    well, she added, that after making a point in class one day, his fellow
    Lutherans took him for a crypto-Catholic! In fact, what I am develop-
    ing in the contrast between prose and version—but the reading would
    be become complicated if I highlighted it every time with quotation
    marks—is that it is a "manner of making prose" within the multiple
    versions of prose.

    [Despret's distinction between "prose" and "version" refers to spe-
    cific terminology in the practice of translation in French. A "prose" is
    when one translates *from* French into a foreign language; a "version"
    is when one translates from a foreign language *into* French. It is this
    terminology that Despret both draws on and reinterprets—gives a
    new version to, if you will—in her writings. "Prose," in this context,
    and as a translation of *thème,* carries this technical, directional sense
    of translation, as well as the dated sense of "to prose," as in to com-
    pose, convert, or translate. This distinction also carries a resonance
    with that of "prose" vs. "verse." I thank Matt Chrulew for drawing
    my attention to this.—Trans.]

5   [The French expression is *forts en thème,* which has an obvious reference
    to "theme" in this context. The expression has a slightly pejorative
    tone, such as to refer to someone as being a "know-it-all."—Trans.]

6   Barbara Cassin, "Relativité de la traduction et relativisme," paper pre-
    sented at La Pluralité Interprétative Colloquium, Collège de France,
    Paris, June 12–13, 2008. Cassin's concerns can also be found in the

engaging book she edited, *Vocabulaire européen des philosophies* (Paris: Le Seuil, 2004). She proposed a version of her experiences in the book I cowrote with Isabelle Stengers, *Women Who Make a Fuss*, trans. April Knutson (Minneapolis, Minn.: Univocal Press, 2014).

7 Eduardo Viveiros de Castro, "Perspectival Anthropology and the Method of Controlled Equivocation," *Tipití: Journal of the Society for the Anthropology of Lowland South America* 2, no. 1 (2004): 5.

8 [See Vinciane Despret, *Penser Comme un Rat* (Versailles: Éditions Quae, 2009). Part of this book has been translated as "Thinking Like a Rat," trans. Jeffrey Bussolini, *Angelaki: Journal of the Theoretical Humanities* 20, no. 2 (2015): 121–34.—Trans.]

9 Viveiros de Castro, "Perspectival Anthropology," 5.

10 The idea that mourning is an impoverished concept for the relation between the living and the dead was raised in a fascinating way in Magali Molinié's book *Soigner les morts pour guérir les vivants* (Paris: Les Empêcheurs de penser en rond, 2003).

11 See Vinciane Despret, *Au bonheur des morts: Récits de ceux qui restent* (Paris: La Découverte/Les empêcheurs de penser en rond, 2015).

12 For other terms denoting dominance and that allow another story to be told, such as Power's use of "charisma," see Margareth Power, *The Egalitarian: Human and Chimpanzee* (Cambridge: Cambridge University Press, 1991); for Zahavi and his use of "prestige," I refer you to "F for Fabricating Science," and for Rowell, to "H for Hierarchies."

13 The question of versions has been important to me for many years now. If it is enriched by these new readings and tests, it could not have been "kept in check" if it were not for the possibilities (the "capturing") opened up by the collective work that led to its elaboration and for which each modification holds a memory (thank you to Didier Demorcy, Marco Mattéos Diaz, and Isabelle Stengers).

14 See Haraway, *When Species Meet*.

W FOR WORK

1 For the work of Jocelyne Porcher cited throughout this chapter, please see *Vivre avec les animaux*. See also Porcher and Tiphaine Schmitt, "Les vaches collaborent-elles au travail? Une question de sociologie," *La Revue du Mauss* 35, no. 1 (2010): 235–61. For additional consultation, see Porcher, *Éleveurs et animaux: réinventer le lien* (Paris: Presses Universitaire de France, 2002), and Porcher and Christine Tribondeau, *Une vie de*

cochon (Paris: Les Empêcheurs de penser en rond / La Découverte, 2008). The critique of industrial systems, as well as the work of observing cows, comes from the article published in *Revue du Mauss*.

2 Richard L. Tapper, "Animality, Humanity, Morality, Society," in *What Is an Animal?*, ed. Timothy Ingold, 47–62 (London: Routledge, 1994). Cited in Porcher, *Vivre avec les animaux*.

3 ["Remember" and "re-member" are in English in the original.—Trans.]

4 See Jérôme Michalon's doctoral dissertation, "L'animal thérapeute: Socio-anthropologie de l'émergence du soin par le contact animalier," presented and defended in Sociology and Political Anthropology, under the direction of Isabelle Mauz, at the Université Jean Monnet de Saint-Étienne, September 2011.

5 See Haraway, *When Species Meet*.

6 See Jocelyne Porcher and Tiphanie Schmitt, "Dairy Cows: Workers in the Shadows?," *Society and Animals* 20 (2012): 39–60.

7 Porcher, *Vivre avec les animaux*, 145.

X FOR XENOGRAFTS

1 [Orson Scott Card, *Speaker for the Dead* (New York: Tor Books, 1986). —Trans.]

2 The scientific article that describes the research of Céline Séveno, Michèle Fellous, Joanna Ashton-Chess, Jean-Paul Soulillou, and Bernard Vanhove on the reconfiguration of Gal-KO has been published as "Les xénogreffes finiront-elles par être acceptées?," *Médecine / Sciences* 21, no. 3 (2005): 302–8.

3 Citations from Catherine Rémy are drawn from her book *La fin des bêtes*. The third section tells the story of her fieldwork within laboratories. I am just as inspired by her more theoretical article on the history of xenografts, "Le cochon est-il l'avenir de l'homme? Les xénogreffes et l'hybridisation du corps humain," *Terrain* 52, no. 1 (2009): 112–25.

4 Haraway, *When Species Meet*, 31.

5 Lynn Margulis and Dorian Sagan, *Acquiring Genomes: A Theory of the Origins of Species* (New York: Basic Books, 2002), 55–56. Cited in Haraway, *When Species Meet*, 31.

6 Haraway, *When Species Meet*, 31; Margulis and Sagan, *Acquiring Genomes*, 205.

7 For the origin of *xenos*, I consulted Pierre Vilard's article "Naissance

d'un mot grec en 1900: Anatole France et les xénophobes," *Les Mots* 8 (1984): 191–95.

8 Séveno et al., "Les xénogreffes finiront-elles par être acceptées?," 306–7.

9 Porcher, *Vivre avec les animaux.*

Y FOR YOUTUBE

1 Immanuel Kant, "Physical Geography," in *Natural Science,* ed. Eric Watkins, trans. Olaf Reinhardt (Cambridge: Cambridge University Press, 2012), 586.

2 ["Tardean imitation" refers to the sociological theory of Gabriel Tarde (1843–1904), in which "imitation" is one of three processes in human society: "invention" is the creation of new ideas, orders, values, and so on; "imitation" is the copying and duplication of what already exists; and "opposition" is when different ideas, values, and so on, come into conflict.—Trans.]

3 See Bruno Latour, "Beware, Your Imagination Leaves Digital Traces," http://www.bruno-latour.fr/sites/default/files/P-129-THES-GB.pdf.

4 Gregg Mitman's analysis can be found in "Pachyderm Personalities: The Media of Science, Politics and Conservation," in *Thinking with Animals,* ed. Lorraine Daston and Gregg Mitman, 173–95 (New York: Columbia University Press, 2005).

5 In terms of how one can "discover science" with the advent of videos, I wish to thank my colleague in anthropology Olivier Servais. He helped me considerably in untangling the subtle and complicated connections between YouTube and scientific writings online. I'm also thankful to Éric Burnard, a journalist with Télévision Suisse Romande, who kindly sent me information on religious and political sites that distribute informational clips on the topic of altruism among animals. I'm also thankful to François Thoreau, a doctoral student in political science at the University of Liège, for sharing with me his well-informed and exciting analyses in this area.

6 ["Monkey," "St. Kitts," and "Drunk" are all in English in the original.—Trans.]

7 The idea that popularization can only become interesting if it endears us to the sciences, and shares with us the emotions, difficulties, and debates of scientists, has been the subject of work by Isabelle Stengers and Bruno Latour. E.g., Stengers, *Cosmopolitics I,* trans. Robert Bononno

(Minneapolis: University of Minnesota Press, 2010); Stengers, *Cosmopolitics II*, trans. Robert Bononno (Minneapolis: University of Minnesota Press, 2011); Latour, *Chroniques d'un amateur de sciences* (Paris: Presses de l'École des mines, 2006).

Z FOR ZOOPHILIA

1   Information on the first zoophile sentenced by the state of Washington following the Pinyan case can be found in the *Seattle Times*, October 20, 2006, http://www.seattletimes.com/seattle-news/spanaway-man-first-arrest-under-states-new-bestiality-law/.

2   See Marcela Iacub and Patrice Maniglier, *Anti-manuel d'éducation sexuelle* (Paris: Bréal, 2007), as well as a number of writings on Iacub's blog, "Être derange avec Marcela Iacub," http://www.culture-et-debats.over-blog.com/.

3   My chapter draws on Brown and Rasmussen's article on queer geography. See Michael Brown and Claire Rasmussen, "Bestiality and the Queering of the Human Animal," *Environment and Planning D: Society and Space* 28 (2010): 158–77.

4   Ibid., 159.

5   Ibid.

6   Michel Foucault, "Sexual Morality and the Law," in *Politics, Philosophy, Culture: Interviews and Other Writings, 1977–1984*, ed. Lawrence D. Kritzman, trans. Alan Sheridan (New York: Routledge, 1990), 281.

7   [Both "think tank" and "intelligent design" are in English in the original. —Trans.]

8   Brown and Rasmussen, "Bestiality and the Queering of the Human Animal," 169.

9   Ibid., 171.

10   Thierry Hoquet, "Zoophilie ou l'amour par-delà la barrier de l'espèce," *Critique* 747–48 (2009): 682. His analysis highlights a different transgression of boundaries (or borders) than my own: those that differentiate, by gender, the meaning of penetration within the human community. He makes note of the conflation, which is often made, between homosexuality and bestiality (and he furthermore shows that the question of consent is but a mask): "It's as if we were dealing here with a lot of sex for no reason between beings deprived of reason."

11   See Rémy, *La fin des bêtes*.

INDEX

(continued from page ii)

Vinciane Despret is associate professor of philosophy at the University of Liège and at the Free University of Brussels. She was scientific curator of the exhibition *Bêtes et Hommes* at the Grande halle de la Villette in Parc de La Villette, Paris, and has collaborated with philosophers, artists, choreographers, filmmakers, and scientists. She is the author or coauthor of nine books, including *Our Emotional Makeup* and *Women Who Make a Fuss* (with Isabelle Stengers; Minnesota, 2014).

Brett Buchanan is director of the School of the Environment and associate professor of philosophy at Laurentian University, Sudbury, Ontario, Canada. He is the author of *Onto-Ethologies: The Animal Environments of Uexküll, Heidegger, Merleau-Ponty, and Deleuze* and coeditor of a special issue of *Angelaki* devoted to the writings of Vinciane Despret.

Bruno Latour is a French philosopher, anthropologist, and sociologist of science.

Made in the USA
Las Vegas, NV
16 January 2022